Annals of Mathematics Studies

Number 78

STRONG RIGIDITY OF LOCALLY SYMMETRIC SPACES

BY

G. D. MOSTOW

PRINCETON UNIVERSITY PRESS
AND
UNIVERSITY OF TOKYO PRESS

PRINCETON NEW JERSEY
1973

LCC: 73–13003
ISBN: 0–691–08136–0

Published in Japan exclusively by
University of Tokyo Press;
In other parts of the world by
Princeton University Press

Printed in the United States of America
by Princeton University Press, Princeton, New Jersey

Library of Congress Cataloging in Publication data will
be found on the last printed page of this book

Contents

STRONG RIGIDITY OF LOCALLY SYMMETRIC SPACES

STRONG RIGIDITY OF LOCALLY SYMMETRIC SPACES

G. D. Mostow[1]

§1. Introduction

The phenomenon of strong rigidity that we establish is the following:

Let Y be a *locally symmetric* Riemannian space; that is, given any point p in Y, the symmetry map $\sigma_p : \exp ty \to \exp -ty$ is an isometry on some ball in Y of center p, where $t \to \exp ty$ denotes the geodesic through p with tangent vector y at p. Assume that the sectional curvature at each point of Y is non-positive. Assume moreover that Y is compact and connected. We prove (cf. Theorem 24.1′)

THEOREM A. *The fundamental group* $\pi_1(Y)$ *determines* Y *uniquely up to an isometry and a choice of normalizing constants, provided that* Y *has no closed one or two dimensional geodesic subspaces which are direct factors locally.*

The normalizing constants referred to in our theorem involves changing the metric of Y in such a way that the symmetry maps σ_p remain unchanged for all $p \in Y$. The constants arise in a canonical fashion. For let X denote the simply connected covering space of Y. Then X is a *symmetric* Riemannian space, that is the symmetry map σ_p is an global isometry of X onto X. The set of symmetries $\{\sigma_p;\ p \in X\}$ generate a group G of isometries which act transitively on X, and the connected

[1] Supported in part by the National Science Foundation Grant GP 33893X.

component of 1 of the group G is a direct product $G^0 = G_0 \times G_1 \times \ldots \times G_n$ where G_0 is its center and $\{G_1, G_2, \ldots, G_n\}$ is the set of all its non-abelian normal, simple, analytic subgroups. Corresponding to this direct product decomposition of G^0, there is a direct product decomposition of $X = X_0 \times X_1 \times \ldots \times X_n$, where X_i is an orbit of G_i, $i = o, \ldots, n$. The space X_0 is a Euclidean space and thus, by hypothesis on Y, reduces to a point. Now each X_i has a unique G_i-invariant metric up to a constant factor c^i for $i = 1, 2, \ldots, n$. These factors $c^1, \ldots, c^n$ are the normalizing constants. In the case that G^0 is simple, they reduce to a single multiplicative constant.

Our theorem can be reformulated as asserting the strong rigidity of discrete subgroups of semi-simple groups. For if we set $\pi_1(Y) = \Gamma$, the group Γ of covering transformations consists of isometries and can be regarded as a subgroup of G^0; we have $Y = \Gamma \backslash X$. Moreover, $X = G/K$ where K is a maximal compact subgroup of G. We call Γ a "co-compact subgroup of G" if and only if G/Γ is compact. Thus, our Theorem is equivalent to

THEOREM A′. *Let* G *be a semi-simple analytic group having no center and no compact normal subgroup other than* (1). *Let* Γ *be a discrete co-compact subgroup of* G. *Then the pair* (G, Γ) *is uniquely determined by* Γ, *provided* PSL(2, R) *is not a direct factor of* G *which is closed modulo* Γ.

That is, given two such pair (G, Γ) *and* (G', Γ') *and an isomorphism* $\theta : \Gamma \to \Gamma'$ *there exists an analytic isomorphism* $\bar{\theta} : G \to G'$ *such that* θ *is the restriction of* $\bar{\theta}$ *to* Γ, *provided there is no factor* G_i *isomorphic to* PSL(2, R) *such that* $\Gamma\, G_i$ *is a closed subgroup of* G.

The reason for the proviso is well-known from uniformization theory. Set $G = PSL(2, R)$. Then G operates on the upper half-plane $X = \{z \in C;\ \mathrm{Im}\ z > 0\}$ via

$$z \to \frac{az + b}{cz + d}$$

and it preserves the metric given by $ds^2 = \frac{dz\,d\bar{z}}{(\mathrm{Im}\ z)^2}$. Given a compact Riemann surface Y of genus greater than one, then its simply connected covering space is analytically equivalent to X. Therefore $\pi_1(Y)$ may be identified with a subgroup Γ of G; clearly Γ is discrete and co-compact.

It is well-known that two compact Riemann surfaces Y and Y′ of the same genus have isomorphic fundamental groups Γ and Γ' but need not be analytically equivalent; that is Γ and Γ' are not in general conjugate under an automorphism of G. Thus strong rigidity fails for $PSL(2,\mathbf{R})$. However, it is the *only* factor causing the failure of strong rigidity for any semi-simple analytic group – or equivalently, for a locally symmetric space.

The chronology of rigidity begins with the theorem of A. Selberg ([16]) that a discrete co-compact subgroup Γ of $SL(n,\mathbf{R})$ cannot be continuously deformed except trivially, that is, by inner automorphisms of $SL(n,\mathbf{R})$, if $n > 2$; Selberg's proof rested on showing that the trace of elements in Γ are preserved under deformations of Γ. Selberg's method applied to the other classical groups of rank greater than 1. At about the same time, E. Calabi and E. Vesentini proved the rigidity of complex structure under infinitesimal deformations of compact quotients of bounded symmetric domains ([3b]), and later Calabi proved the metric analogue for compact hyperbolic n-space forms for $n > 2$ ([3a]). Thereupon A. Weil ([21]) generalized Selberg's and Calabi's results to semi-simple groups having no compact or 3 dimensional simple factors. Weil's proof deduces the rigidity of Γ in G under deformations from the vanishing of the cohomology group $H^1(\Gamma,\dot{G})$ where $\dot{G}$ is the Lie algebra of $\dot{G}$ regarded as a Γ-module under the adjoint representation. In the case of arithmetic subgroups, which are lattices (that is, Γ is discrete in G and G/Γ has finite Haar measure) but not generally co-compact, the rigidity under deformations was proved independently by A. Borel (unpublished) in the case of $\mathbf{Q}$-simple groups of $\mathbf{Q}$-rank at least two, and H. Garland (in the split case), and by M. S. Raghunathan in the remaining $\mathbf{Q}$-rank one cases, again by showing that $H^1(\Gamma,\dot{G}) = 0$.

The phenomenon of strong rigidity for arbitrary lattices first turned up in 1965 in my search for a geometric explanation of deformation-rigidity.

The point of view adopted here can be explained in terms of the counterexample cited above. When we regard the fundamental groups of two complete Riemann surfaces Y and Y' of the same genus p and of finite volume as subgroups Γ and Γ' of the isometry group of the upper half plane X, they are isomorphic transformation groups; that is, there is a diffeomorphism of Y onto Y' and its lift provides a Γ-space isomorphism $\phi : X \to X'$ such that for all $\gamma \in \Gamma$ and $x \in X$,

$$\phi(\gamma x) = \theta(\gamma)\, \phi(x)$$

where $\theta : \Gamma \to \Gamma'$ is given by the isomorphism of $\pi_1(Y)$ to $\pi_1(Y')$. As transformation groups, what distinguishes Γ from Γ'? The answer is: On X they are indistinguishable. But on $X \cup X_0$, where X_0 is the boundary of X, Γ is conjugate to Γ' in G according as the Γ-space morphism ϕ is *smooth* at X_0 or not. Moreover this result is true *without exceptions* for any semi-simple analytic group having no compact normal subgroups. (cf. [12f]; also [12i] for a more detailed account.)

In view of the known deformation-rigidity theorems, it was natural to conjecture that the hypotheses of *existence of boundary values* and *smoothness at the boundary* are superfluous for all symmetric spaces of negative curvature having no two-dimensional factors. This conjecture was confirmed in [12g] for the case of two diffeomorphic compact spaces of constant negative curvature of dimension exceeding two. The proof relied heavily on analytic tools – not surprisingly in view of the analytic nature of the problem. Fortunately, the theory of quasi-conformal mappings in 3-space was at hand; upon generalizing the theory to n-space, one could be certain that the map ϕ took on continuous boundary values which were almost everywhere differentiable. Utilizing the ergodic action of Γ at infinity, I could prove that the boundary map ϕ_0 was actually a Möbius transformation if $n > 2$.

The relevance of the usual theory of quasi-conformal mappings is unfortunately limited to spaces of constant curvature only. For the case of arbitrary symmetric spaces, it was necessary to find an entirely different method to establish that ϕ takes on boundary values.

The method adopted here relies on the key notion of a *pseudo-isometry*, which is defined as follows:

Let $k \geq 1$ and $b \geq 0$. A map $\phi : X \to X'$ between metric spaces is called a (k, b) pseudo-isometry if

(1) $$d(\phi(x), \phi(y)) \leq k\, d(x, y) , \qquad x, y \in X$$

(2) $$d(\phi(x), \phi(y)) \geq k^{-1}\, d(x, y) , \qquad \text{for all } x, y \text{ in}$$
X such that $d(x, y) \geq b$, where d denotes distance.

A map $\phi : X \to X'$ is called a pseudo-isometry if it is a (k, b) pseudo-isometry for some (k, b).

Our proof of strong rigidity consists of four main steps.

(i) Let Y and Y' be as in Theorem A, let X and X' denote their simply connected covering spaces, and let $\pi_1(Y) = \Gamma = \pi_1(Y')$. Then there is a Γ-space pseudo-isometry $\phi : X \to X'$ (ϕ need not be injective).

(ii) The Γ-space pseudo-isometry ϕ has continuous boundary values ϕ_0.

(iii) The map ϕ_0 induces an incidence preserving isomorphism of the "Tits geometry" of G onto the Tits geometry of G'. If X has flat geodesic subspaces of dimension greater than 1, then the Generalized Fundamental Theorem of Projective Geometry shows that ϕ_0 is induced by an analytic isomorphism of G to G'.

(iv) If X has no flat geodesic subspace of dimension exceeding 1, then $X = H^n_{\mathbf{K}}$, hyperbolic kn-space over the division algebra $\mathbf{K} = \mathbf{R}, \mathbf{C}, \mathbf{H}$ (quaternions) or $\mathbf{O}$ (Cayley numbers), where $k = \dim_{\mathbf{R}} \mathbf{K}$.

In this case the usual theory of quasi-conformal mappings does not work for $\mathbf{C}$, $\mathbf{H}$, and $\mathbf{O}$ as it does for $\mathbf{R}$. However, we introduce the notion of a $\mathbf{K}$-*quasi-conformal mapping over a division algebra* $\mathbf{K}$, which amounts to the usual theory if $K = R$. We then generalize some aspects of the usual quasi-conformal mapping theory to obtain the absolute continuity of the boundary map ϕ_0 along almost all "R-circles" (cf. Section 21). Again, as in the case of H^n_R, the ergodicity of Γ at infinity allows us to prove that the boundary map ϕ_0 is induced by an isomorphism of G to G'.

The proof presented here is an implementation of the program that was first formulated in 1965 and announced in [12f]. The successful execution of our strategy for the case of no R-rank 1 factors was announced at the International Congress of 1970 (cf. [12j]).

During that intervening period, strong rigidity had been proved for certain arithmetic (non co-compact) subgroups on the one hand in 1967 by Bass-Milnor-Serre (cf. [1]) and on the other by M. S. Raghunathan [14], using algebraic and arithmetic methods.

Also, G. A. Margulis has announced results stating that for R-rank > 1, every non co-compact irreducible lattice is nearly arithmetic (cf. [10]) and in 1971 Margulis announced strong rigidity for such non co-compact irreducible lattices.

It is appropriate to mention prior partial results on the extent to which Γ determines G. In 1967 H. Furstenberg studied this question from a probabilistic point of view and was led to interesting questions about the Poisson boundary (cf. [6b]); his method yielded that a lattice in $SL(n, R)$ $(n \geq 3)$ could not be a lattice in $SO(1, n)$. In 1962 J. Wolf pointed out that if G/Γ is compact, the *rank* of the symmetric space associated to G was determined by Γ (cf. [22]).

Although we have stated Theorem A′ for co-compact lattices only, our method applies for most part to arbitrary lattices. Indeed we require the co-compactness of Γ only to assure the *existence* of a Γ-space pseudo-isometry $\phi: X \to X'$. In those cases such as lattices in R-rank one groups or arithmetic lattices where information is available about the cusps of a fundamental domain for Γ in X, it is possible to prove the existence of a Γ-space pseudo-isometry ϕ. Indeed the existence of such a ϕ has been recently proved by Gopal Prasad for the case of $\mathbf{R}$-rank one spaces; this establishes strong rigidity for non-compact lattices in the $\mathbf{R}$-rank one cases other than $PSL(2, \mathbf{R})$

Thus combining our results with Margulis' and Prasad's, one can weaken the hypothesis of co-compactness of Γ in Theorem A to: "Let Γ be a lattice in G." That is to say, *strong rigidity* holds for *arbitrary lattices* in centerless analytic semi-simple groups without compact factors apart from the aforementioned exception.

In Theorem A this amounts to assuming merely that Y has finite measure rather than Y is compact.

In conclusion, some sections of our proof have independent interest. The theory of pseudo-isometries which plays such a central role in our method may be useful in other contexts (cf. Sections 9, 10, 12). The theory of quasi-conformal mappings over the division algebra $\mathbf{K}$ ($\mathbf{K} = \mathbf{R}$, $\mathbf{C}$, $\mathbf{H}$, or $\mathbf{O}$) which intervenes implicitly in Sections 20, 21 deserves further attention.

§2. Algebraic Preliminaries

2.1. We denote by $M(n, \mathbb{R})$ the set of all $n \times n$ real matrices. We denote by $GL(n, \mathbb{R})$ the group of invertible $n \times n$ real matrices. A subgroup G of $GL(n, \mathbb{R})$ is called *algebraic* if G is the set of common zeroes in $GL(n, \mathbb{R})$ of $P_1, \ldots, P_m$ where each P_i is a polynomial in the matrix coefficients with real coefficients. An element of $GL(n, \mathbb{R})$ is called *semi-simple* if its minimal polynomial has no repeated factors – or equivalently, if it can be diagonalized over the complex numbers. An element u is called *unipotent* if $(u-1)^n = 0$; i.e., all its eigenvalues are 1. The *Jordan normal form* for elements of $GL(n, \mathbb{R})$ implies: Given $g \in GL(n, \mathbb{R})$, there are elements s and u in $GL(n, \mathbb{R})$ satisfying: $g = su$, $su = us$, s is semi-simple, u is unipotent. Moreover, these conditions determine s and u uniquely. One calls s and u the semi-simple and unipotent *Jordan components* respectively. Any element commuting with g commutes with its Jordan components.

Let s be a semi-simple element in $GL(n, \mathbb{R})$. Then by suitably pairing the eigenspaces of s over the complex numbers, one can find semi-simple elements k and p in $GL(n, \mathbb{R})$ satisfying: $s = kp$, $kp = pk$, the eigenvalues of k have modulus 1, the eigenvalues of p are positive. Moreover, these conditions determine k and p uniquely, and any element commuting with s commutes with k and p. We call p the *polar part* of s. For any $g \in GL(n, \mathbb{R})$ we denote by *pol* g the polar part of the Jordan semi-simple component of g.

If G is an algebraic group, and $g \in G$, then G contains the Jordan components of g (cf. [4]) and *pol* g (cf. [12a]), as well as $u^t = \exp t \log u$ for all $t \in \mathbb{R}$, where u is the unipotent Jordan component of g. If $g = \mathit{pol}\, g$, then for any real number t, $g^t = \exp t \log g$ is a well defined semi-simple element with positive eigenvalues in $GL(n, \mathbf{R})$ and belongs to every algebraic group containing g.

An arcwise connected abelian subgroup A of $GL(n, \mathbf{R})$ such that $g = pol\, g$ for all $g \in A$ is called a *polar subgroup*. Any polar subgroup is isomorphic to $\mathbf{R}^s$ for some s.

If V is a finite dimensional vector space over $\mathbf{R}$, then upon identifying $\mathrm{Hom}_{\mathbf{R}}(V, V)$ and $\mathrm{Aut}\, V$ with $M(n, \mathbf{R})$ and $GL(n, \mathbf{R})$ via a choice of base, one can define algebraic subgroups and polar subgroups of $\mathrm{Aut}\, V$ in an unambiguous way. Clearly any polar subgroup of $\mathrm{Aut}\, V$ can be put in diagonal form via a suitable choice of base in V.

2.2. Let G be a Lie subgroup of $GL(n, \mathbf{R})$. We denote by $\dot{G}$ the Lie algebra of G, and we identify $\dot{G}$ with $\{Y;\ Y \in M(n, \mathbf{R}),\ \exp \mathbf{R}Y \subset G\}$; that is, with the tangent space to G at the identity element 1. The *adjoint representation* is given by $\mathrm{Ad}\, g(Y) = gYg^{-1}$, where $g \in G$, $Y \in \dot{G}$.

Let A be a polar subgroup of the Lie subgroup G. Then $\mathrm{Ad}\, A$ can be diagonalized and therefore

$$\dot{G} = \sum \dot{G}_\alpha \qquad \text{(direct)}$$

where each α is an analytic homomorphism of A into the multiplicative group of positive real numbers and $\dot{G}_\alpha = \{Y \in \dot{G};\ \mathrm{Ad}\, g(Y) = \alpha(g)\, Y\}$. The set of α such that $\dot{G}_\alpha \neq 0$ are called $\mathbf{R}$-roots of G on A. Each subspace $\dot{G}_\alpha$ is stable under $\mathrm{Ad}\, Z(A)$ where $Z(A)$ denotes the centralizer of A in G. If α and β are roots, then $[\dot{G}_\alpha, \dot{G}_\beta] \subset \dot{G}_{\alpha\beta}$. Moreover, each element in $\dot{G}_\alpha$ is *nilpotent* if $\alpha \neq 1$.

The maximal polar subgroups of G are conjugate via inner automorphisms [12d].

2.3. Let G be an analytic subgroup of $GL(n, \mathbf{R})$; that is, G is a connected Lie subgroup. Assume moreover that G is *semi-simple*; that is, G has no normal analytic abelian subgroup other than (1). Since G is its own commutator subgroup, each element of G has determinant one.

Let A be a maximal polar subgroup of G, and let χ denote the group of analytic homomorphisms of A into R^+, the multiplicative group of positive real numbers. Then χ is an abelian vector group. Writing the multiplication in χ additively, we have $(\alpha+\beta)(a) = \alpha(a)\beta(a)$ for $a \in A$.

The R-roots of G on A satisfy almost all of the well-known properties of root systems on Cartan subalgebras of complex semi-simple Lie algebras. Namely

(i) If α is an R-root, then $-\alpha$ is an R-root.

(ii) If $\alpha, \beta, \alpha+\beta$ are R-roots, then $[\dot{G}_\alpha, \dot{G}_\beta] \neq 0$.

(iii) There is a subset of linearly independent R-roots $\alpha_1, \dots, \alpha_r$ such that any root has the form $\pm \Sigma_i n_i \alpha_i$ with n_i non-negative integers. A subset $\alpha_1, \dots, \alpha_r$ with this property is called a *fundamental system* of R-roots.

(iv) Let N(A) denote the normalizer of A in G and let Z(A) denote the centralizer of A in G. Then N(A)/Z(A) operates (via inner automorphisms) simply-transitively on the set of fundamental systems of R-roots.

(v) The R-roots separate the points of A. Thus $A \approx R^r$.

A semi-simple analytic group G is the topologically connected component of the identity of the smallest algebraic group G^* in GL(n, R) containing G. Therefore for any $g \in G$, the semi-simple group G contains the one parameter groups $(pol\ g)^t$ and u^t $(t \in R)$ where u is the unipotent Jordan component of g. In particular, *pol* g lies in a maximal polar subgroup of G and $u \in G$.

DEFINITION. An element g in G is called *polar regular* if

$$\dim Z\,(pol\ g) \leq \dim Z\,(pol\ h)$$

for all $h \in G$, where Z(h) denotes the centralizer of h in G.

Let g be an element of G with $g = pol\ g$. We call g R-regular if g is polar regular, and R-singular if g is not polar regular.

Let A be a maximal polar subgroup of the semi-simple analytic group G. Inasmuch as all the maximal polar subgroups are conjugate in G, A contains R-regular elements. Indeed an element $a \in A$ is R-singular if and only if $\alpha(a) = 1$ for some R-root α where α is not the zero element of χ. An element $g \in A$ is R-regular if and only if $Z(g) = Z(A)$. An R-regular element lies in a unique maximal polar subgroup.

DEFINITION. dim A is called the R-rank of G.

Let $\alpha_1, \ldots, \alpha_r$ be a fundamental system of R-roots on A. Set

$$◂A = \{a \in A;\ \alpha_i(a) > 1,\quad i = 1, 2, \ldots, r\}$$

Then

(vi) Any R-regular element of A is conjugate to an element of $◂A$ via an inner automorphism from N(A).

From (vi) it follows at once that the orbit of $◂A$ under N(A) is the set $A = \bigcup_{\alpha \neq 0} \operatorname{Ker} \alpha$. Inasmuch as $◂A$ is clearly a topologically connected component of $A - \bigcup_{\alpha \neq 0} \operatorname{Ker} \alpha$, we see that the fundamental systems of R-roots are in one-to-one correspondence with the topologically connected components of $A - \bigcup_{\alpha \neq 0} \operatorname{Ker} \alpha$. We call these subsets of A *chambers*. Each chamber is a convex cone with r-faces of codimension 1 and with $\binom{r}{i}$ faces of codimension i. A face of a chamber is called a *chamber wall*. In particular, the identity element constitutes a wall; we call it the "0-face" of the chamber.

(vii) The topological closure $◂\bar{A}$ of a chamber $◂A$ is a fundamental domain in the strictest sense for the action of the group N(A)/Z(A) on A; that is, if $g \in N(A)$, $a \in ◂\bar{A}$, and $gag^{-1} \in ◂\bar{A}$, then $gag^{-1} = a$. By 2.2, the maximal polar subgroups of G are conjugate under an inner automorphism. It follows that if $g \in G$, $a \in ◂\bar{A}$, and $gag^{-1} \in ◂\bar{A}$, then $gag^{-1} = a$ and $g \in Z(a)$.

2.4. We continue the assumptions and notations of 2.3.

Let $◂B$ be a chamber or chamber wall in A. For any R-root α we write $\alpha(◂B) > 0$ if $\alpha(a) > 1$ for all $a \in ◂B$; $\alpha(◂B) = 0$ if $\alpha(a) = 1$ for all $a \in ◂B$.

The subspace $\Sigma_\alpha \dot{G}_\alpha$, $\{\alpha; \alpha(\blacktriangleleft B) > 0\}$, is a Lie subalgebra of nilpotent elements stable under inner automorphisms from $Z(B)$ (cf. 2.3). Let $U(\blacktriangleleft B)$ denote the corresponding analytic group. Then $U(\blacktriangleleft B)$ is a group of unipotent elements and is in fact unipotent, it is conjugate in $GL(n, \mathbf{R})$ to a group of triangular matrices. Set

$$P(\blacktriangleleft B) = Z(\blacktriangleleft B)\, U(\blacktriangleleft B)$$

Then $P(\blacktriangleleft B)$ is a group, $U(\blacktriangleleft B)$ is normal in $P(\blacktriangleleft B)$ and $P(\blacktriangleleft B) = N(U(\blacktriangleleft B)) = N(P(\blacktriangleleft B))$, where $N(\ )$ denotes the normalizer of $(\ )$ in G. From $P(\blacktriangleleft B) = N(P(\blacktriangleleft B))$ it follows that $P(\blacktriangleleft B)$ is closed in G.

The unipotent group $U(\blacktriangleleft A)$ is a *maximal* unipotent subgroup of G for any *chamber* $\blacktriangleleft A$ and moreover all maximal unipotent subgroups of G are conjugate via an inner automorphism of G [12d].

If $\blacktriangleleft B_1$ and $\blacktriangleleft B_2$ are chambers or chamber walls, then $P(\blacktriangleleft B_1) \subset P(\blacktriangleleft B_2)$ if and only if $\blacktriangleleft B_2$ is a face of $\blacktriangleleft B_1$. The subgroup $P(\blacktriangleleft B)$ is called the *parabolic* subgroup associated to $\blacktriangleleft B$. $P(\blacktriangleleft B)$ is a *minimal* parabolic subgroup if and only if $\blacktriangleleft B$ is a chamber. Corresponding to the conjugacy of all the chambers in G, we have the conjugacy in G of all the minimal parabolic subgroups.

2.5. For any topological group G we denote by G^0 the topologically connected component of the identity. If G is an algebraic group, then G/G^0 is finite [12c]. For any subset $S \subset GL(n, \mathbf{R})$, $Z(S)$ is an algebraic group and therefore $Z(S) \cap G^0$ has only a finite number of connected topological components. In particular $P(\blacktriangleleft B)$ has only a finite number of topologically connected components for any chamber wall $\blacktriangleleft B$.

2.6. Let V be a finite dimensional vector space over $\mathbf{R}$ and let G be a semi-simple analytic subgroup of Aut V. Then G is self-adjoint with respect to some positive definite quadratic form on V [12b]. Selecting an orthonormal base with respect to the form, and identifying Aut V with $GL(n, \mathbf{R})$, we find ${}^tG = G$, where ${}^tg$ is the transpose of the matrix g. Set

$$P(n, R) = \{x \in GL(n, R);\ {}^t x = x,\ x \text{ positive definite}\}\ .$$
$$O(n, R) = \{x \in GL(n, R);\ {}^t x = x^{-1}\}\ .$$
$$S(n, R) = \{x \in M(n, R);\ {}^t x = x\}\ .$$

As is well-known $GL(n, R) = P(n, R) \cdot O(n, R)$ and $\exp : S(n, R) \to P(n, R)$ is a bianalytic homeomorphism.

For any $g \in G$, the positive definite symmetric element $g^t g$ lies in G and hence by 2.3, $(g^t g)^{\frac{1}{2}} \in G \cap P(n, R)$. It follows at once that

(i) $G = (G \cap P(n, R)) \cdot (G \cap O(n, R))$

this decomposition being a direct product topologically. Since G is closed in $GL(n, R)$,

(ii) $G \cap O(n, R)$ is a compact subgroup of G.

Moreover (cf. [12b])

(iii) Any compact subgroup of G is conjugate via an inner automorphism to a subgroup of $G \cap O(n, R)$.

In fact properties (i), (ii), and (iii) are valid for any self-adjoint group which is of finite index in an algebraic group. Thus if $S \subset G \cap O(n, R)$ or $S \subset G \cap P(n, R)$, then ${}^t Z(S) = Z(S)$ and

$$Z(S) = (Z(S) \cap P(n, R)) \cdot (Z(S) \cap O(n, R))$$

We consider two special cases of this observation. Set $K = G \cap O(n, R)$. Let A be a maximal abelian subgroup of $P(n, R)$. By definition, $Z(A) \cap P(n, R) = A$ and therefore by (i)

(iv) $Z(A) = (Z(A) \cap K) \cdot A$.

This implies that A is a maximal polar subgroup of G.

As a second case, let s be a semi-simple element of G and let $k \cdot p = s$ be its polar decomposition with $p = pol\ s$. We wish to show that $Z(s)$ is conjugate to a self-adjoint group. By 2.2, p is conjugate to an element of A. Without loss of generality we can assume that $p \in G \cap P(n, R)$. Hence ${}^t Z(p) = Z(p)$ and $Z(p) = (Z(p) \cap P(n, R) \cdot (Z(p) \cap O(n, R))$. Since k lies in a compact subgroup of $Z(p)$, we can assume after conjugation by an element of $Z(p)$ that $k \in Z(p) \cap O(n, R)$, by (iii). Thus $Z(s) = Z(k) \cap Z(p) = {}^tZ(s)$. Thus we see that $Z(s)$ operates semi-simple ("reductively") on the underlying vector space for any semi-simple element s of the semi-simple group G. Such a group is called *reductive* and its Lie algebra is easily seen to be the direct sum of its center and a semi-simple ideal.

From the arguments used above one deduces easily

(v) Any elements of $G \cap P(n, \mathbf{R})$ that are conjugate in G are conjugate by an element of $G \cap O(n, \mathbf{R})$.

2.7. Let G, A, K be as above. For any $\mathbf{R}$-root α on the maximal polar subgroup A, we have

$$a\, X_\alpha\, a^{-1} = \alpha(a)\, X_\alpha$$

$$a^{-1}\, {}^tX_\alpha\, a = \alpha(a)\, {}^tX_\alpha$$

for all $a \in A$, $X_\alpha \in \dot{G}_\alpha$. Thus ${}^tX_\alpha \in \dot{G}_{-\alpha}$. Inasmuch as $X_\alpha - {}^tX_\alpha \in \dot{K}$, we get $\dot{G} = \dot{K} + \dot{P}(\blacktriangleleft B)$ for any chamber or chamber wall $\blacktriangleleft B$ in A. That is, $G = KP(\blacktriangleleft B)$ and $G/P(\blacktriangleleft B)$ is compact. Since all chambers in G are conjugate, we see that G/P is compact for any parabolic subgroup P.

REMARK 1. The notion of parabolic subgroup can be defined for algebraic groups G' over an arbitrary field k: An algebraic subgroup P' of G' is parabolic if and only if G'/P' is a *complete* variety. In this context, the points of G are taken with coordinates in an algebraically closed field containing k. In our situation, let G' be the smallest algebraic group in $GL(n, \mathbf{C})$ containing the semi-simple analytic subgroup $G \subset GL(n, \mathbf{R})$. Let P' denote the smallest algebraic subgroup in $GL(n, \mathbf{C})$ containing $P(\blacktriangleleft B)$. Then P' is a parabolic subgroup of G'. The group P' is necessarily connected and $P(\blacktriangleleft B) = P'_{\mathbf{R}}$. Thus $P'_{\mathbf{R}}$ is connected in the Zariski topology but not necessarily in the Euclidean topology. One can characterize a parabolic subgroup of G as a Zariski-closed subgroup P of G such that G'/P' is compact.

2.8. Let G be a semi-simple analytic group, let A be a maximal polar subgroup of G, let $^{\blacktriangleleft}A$ be a chamber in A, and let $U = U(^{\blacktriangleleft}A)$. The *Bruhat decomposition* of G asserts $G = U\,N(A)\,U$. As a consequence $G = U(^{\blacktriangleleft}B)\,N(A)\,P(^{\blacktriangleleft}B)$ for any chamber wall $^{\blacktriangleleft}B$ in A.

2.9. Let $\mathfrak{G}$ be a semi-simple Lie algebra over a field of characteristic zero and let n be an element of $\mathfrak{G}$ such that ad n is nilpotent. The Jacobson-Morozov lemma says: $\mathfrak{G}$ contains elements h and $n\_$ such that

$$[h,n] = 2n,\ [h,n\_] = -2n\_,\ [n,n\_] = h\ .$$

As a direct consequence, any (non-abelian) reductive analytic group in $GL(n,\mathbb{R})$ which contains a unipotent element $u(u \neq 1)$ contains a three dimensional analytic group with Lie algebra generators h, n, $n\_$ as above and with $u = \exp n$.

2.10. For any symmetric matrix X and skew-symmetric matrix Y we have $\mathrm{Tr}\,XY = -\mathrm{Tr}\,YX = 0$. Hence $S(n,\mathbb{R})$, the tangent space to $P(n,\mathbb{R})$ at 1, is orthogonal to the Lie algebra of $O(n,\mathbb{R})$ with respect to $\mathrm{Tr}\,XY$, and a fortiori the Lie algebra of $G \cap O(n,\mathbb{R})$ is orthogonal to the tangent space to $G \cap P(n,\mathbb{R})$ at 1. This is true for *any* representation of the semi-simple analytic group. Thus one can characterize $\dot{G} \cap S(n,\mathbb{R})$ as the orthogonal complement to $\dot{K}$ with respect to $\mathrm{Tr}\,\mathrm{ad}\,X\,\mathrm{ad}\,Y$.

Set $E = \dot{G} \cap S(n,\mathbb{R})$. The automorphism $x \to {}^t x^{-1}$ of $GL(n,\mathbb{R})$ stabilizes the self-adjoint group G and induces on it an automorphism σ of order 2 such that

$$\dot{\sigma}(X) = X \quad \text{for} \quad X \in \dot{K}$$
$$\dot{\sigma}(X) = -X \quad \text{for} \quad X \in E\ .$$

σ is called the Cartan involution with respect to K. The group $G^1 = \sigma G \cup G$ with multiplication defined by

$$(\sigma g)\,g_1 = \sigma(g\,g_1)$$
$$(\sigma g)(\sigma g_1) = \sigma(g)\,g_1$$

is an extension of order 2 of G having as a maximal compact subgroup $K^1 = \sigma K \cup K$. The injection $G/K \to G^1/K^1$ is bijective.

2.11. Let G be a semi-simple analytic group having a faithful matrix representation and let K be a maximal compact subgroup of G. Set $X = G/K$. Let $\rho : G \to \operatorname{Aut} V$ be a faithful representation of G, V being an n-dimensional vector space over $\mathbf{R}$. Then one can select a base in V with respect to which $\rho(G) = {}^t\rho(G)$ and $\rho(G) \cap O(n, \mathbf{R}) = \rho(K)$ by 2.6 (iii). We have

$$\rho(G) = (\rho(G) \cap P(n, \mathbf{R})) \cdot \rho(K) \ .$$

Let $\mu : X \to P(n, \mathbf{R})$ be the map given by

$$\mu(x K) = \rho(x)\, {}^t\rho(x) \ .$$

Let $[S]$ denote the projective space of lines through the origin of the linear space $S(n, \mathbf{R})$ and let $\pi : S(n, \mathbf{R}) - 0 \to [S]$ denote the the canonical projection. Inasmuch as $\det \mu(x) = 1$ for all $x \in G$, we see that $\pi \circ \mu$ is injective. For any $g \in GL(n, \mathbf{R})$ and for any $y \in S(n, \mathbf{R})$, the canonical operation $y \to g y\, {}^t g$ of $GL(n, \mathbf{R})$ on $S(n, \mathbf{R})$ is linear in y and therefore induces a canonical action of $GL(n, \mathbf{R})$ on the projective space $[S]$. Since $\mu(g \times K) = \rho(g)\, \mu(x K)\, {}^t\rho(g)$ for all $g, x \in G$, we see that the map $\pi \circ \mu$ is a morphism of the G-space X into the $GL(n, \mathbf{R})$-space $[S]$. Let $\overline{X}$ denote the topological closure of $\pi \circ \mu(X)$ in $[S]$. Then $\overline{X}$ is a compact G-space, the *Satake ρ-compactification* of X [15]. Satake has shown that $\overline{X}$ is a finite union of G-orbits, each of dimension less than $\dim X$. Among these G-orbits, there is a unique *compact* G-orbit; we denote this orbit by X_0. The orbit X_0 may be equally well characterized as the G-orbit in $\overline{X}$ of lowest dimension. The isotropy subgroup of a point in X_0 is a parabolic subgroup of G. The Satake ρ-compactification depends on the representation ρ. For suitable ρ (for example, if the "highest $\mathbf{R}$-weight" of ρ lies inside a chamber) then the isotropy subgroup of a point in X_0 is a *minimal* parabolic subgroup. For such a compactification we call X_0 the *Furstenberg maximal boundary* (cf. A Poisson formula for semi-simple groups [6].) Thus for the maximal boundary, $X_0 = G/P(^{\blacktriangleleft}A)$ for some chamber $^{\blacktriangleleft}A$.

We shall give in Section 4 below an alternative definition of X_0 in terms of the metric properties of X.

§3. The Geometry of X: Preliminaries

Let G be a semi-simple analytic group having a faithful matrix representation. Let K be a maximal compact subgroup of G. Set $X = G/K$. One can define a metric on X invariant under G in the following way. Choose a faithful matrix representation ρ (cf. 2.6) such that ${}^t\rho(G) = \rho(G)$ and $\rho(K) = G \cap O(n, \mathbf{R})$. Then imbed X in $P(n, \mathbf{R})$ via the map μ:

$$\mu(xK) = \rho(x)\, {}^t\rho(x)$$

for $x \in G$.

On the space $P(n, \mathbf{R})$, the metric

$$\left(\frac{ds}{dt}\right)^2 = \operatorname{Tr}(p^{-1}\dot{p})^2 ,$$

where $p(t)$ is a differentiable path in $P(n, \mathbf{R})$ is clearly invariant under the canonical $GL(n, \mathbf{R})$ action $y \to g\, y\, {}^tg$. Since $\mu(gxK) = \rho(g)\mu(xK){}^t\rho(g)$, the induced metric on X is G-invariant.

REMARK 1. If G is a simple analytic group, any G-invariant metric on X is unique up to a constant factor. If $G_1, \ldots, G_s$ are the simple normal analytic subgroups of G, then $X = X_1 \times \ldots \times X_s$ (direct) where $X_i = G_i/K \cap G_i$. The most general G-invariant metric on X is the direct product metric and is therefore unique up to multiplication by a constant c_i in each factor X_i, $i = 1, \ldots, s$.

Let $Z = \bigcap_g gKg^{-1}$, $\{g \in G\}$. Then Z is the maximum normal compact subgroup of G and G/Z operates faithfully on X. The group G/Z has no compact normal subgroup and therefore, as one may deduce easily, has no center. Conversely, a semi-simple analytic group having no compact direct factors and no center has no compact normal subgroup other than (1) and thus operates faithfully on X.

REMARK 2. Let τ denote the canonical map of G into Isom X, the group of isometries of X with respect to a G-invariant metric. Then $\tau(G)$ is the connected component of the identity of Isom X and therefore of finite index in it [7]. We shall not require this fact however.

For each point p of the Riemannian space X, there is an isometry σ_p of X such that $\sigma_p(p) = p$ and σ_p sends each tangent vector to X at p into its negative. To see this, it suffices to prove it for a single point of X since G operates transitively on X. For the point $p = 1 \cdot K$, the map σ_p is induced by the map $x \to {}^t x^{-1}$ of $P(n, \mathbf{R})$. The space X is thus a *symmetric Riemannian space*. Its geometry was first studied by E. Cartan. The account of Cartan's results (through Lemma 3.4) presented in this section is taken largely from [12b] and we include the proofs of the lemmas for the sake of expository completeness.

For any elements Y and Z in $M(n, \mathbf{R})$ set

$$\tau_Y(Z) = \frac{d}{dt}(\exp - Y/2 \cdot \exp(Y + tZ) \cdot \exp - Y/2)_{t=0} \cdot$$

For any $Y \in M(n, \mathbf{R})$, let L_Y and R_Y denote the linear maps of $M(n, \mathbf{R})$ given by

$$L_Y(Z) = YZ \ , \qquad R_Y(Z) = ZY$$

for $Z \in M(n, \mathbf{R})$. Set $D_Y = L_Y - R_Y$. On $M(n, \mathbf{R})$ we introduce the inner product

$$\langle U, V \rangle = \operatorname{Tr} U\, {}^t V \ .$$

Then for any $Y \in S(n, \mathbf{R})$, we have

$$\operatorname{Tr}\, [Y, U]\, {}^t V = -\operatorname{Tr} U[Y, {}^t V] = \operatorname{Tr} U^t [Y, V]$$

where $[Y, U] = YU - UY = D_Y U$; thus D_Y is self-adjoint with respect to the inner product.

LEMMA 3.1. *For any* $Y \in S(n, R)$,

$$\tau_Y = \frac{\sinh D_Y/2}{D_Y/2} .$$

Proof. Set $Y_t = Y + tZ$, and denote differentiation by $(\dot{\ })$. We have

$$Y_t \cdot \exp Y_t = (\exp Y_t) Y_t$$

$$\dot{Y}_t \exp Y_t + Y_t (\exp \dot{Y}_t) = (\exp \dot{Y}_t) Y_t + (\exp Y_t) \dot{Y}_t$$

$$Y (\exp \dot{Y})_0 - (\exp \dot{Y})_0 Y = (\exp Y) Z - Z \exp Y$$

where $(\exp Y)_0$ denotes the derivative at $t = 0$. Multiplying on the left and right by $\exp - Y/2$ we get

$$\begin{aligned} D_Y \tau_Y(Z) &= ((\exp - R_Y/2)(\exp L_Y/2) - (\exp R_Y/2)(\exp - L_Y/2))(Z) \\ &= (\exp D_Y/2 - \exp - D_Y/2)(Z) . \end{aligned}$$

Since D_Y is self-adjoint for $Y \in S(n, R)$, we may conclude (cf. [12b])

$$\tau_Y = \frac{\exp D_Y/2 - \exp - D_Y/2}{D_Y} .$$

This establishes Lemma 3.1.

We denote by log the inverse of the exponential map of $S(n, R)$ onto $P(n, R)$.

LEMMA 3.2. *Along any differentiable path* $p(t)$ *in* $P(n, R)$,

$$\mathrm{Tr}\, (\log \dot{p})^2 \leq \mathrm{Tr}\, (p^{-1} \dot{p})^2$$

with equality if and only if p *and* $\dot{p}$ *commute.*

Proof. $\operatorname{Tr}(p^{-1}\dot{p})^2 = \operatorname{Tr}\left(p^{-\frac{1}{2}}\dot{p}\,p^{-\frac{1}{2}}\right)^2$. Set $Y(t) = \log p(t)$. The lemma asserts

$$\operatorname{Tr}\dot{Y}^2 \leq \operatorname{Tr}\tau_Y(\dot{Y})^2$$

with equality if and only if Y and $\dot{Y}$ commute. This follows at once from the fact that $\frac{\sinh D_Y}{D_Y}$ is self-adjoint for each Y and has eigenvalues $\frac{\sinh(\lambda_i-\lambda_j)}{\lambda_i-\lambda_j}$ where $\lambda_1,\ldots,\lambda_n$ are the eigenvalues of Y, together with the fact that $\frac{\sinh\lambda}{\lambda} \geq 1$ for $\lambda \in \mathbf{R}$.

The foregoing lemma has an interesting interpretation. If we define the metric d_S on $P(n,\mathbf{R})$ by the formula

$$\left(\frac{ds}{dt}\right)^2 = \operatorname{Tr}(\dot{\log p})^2$$

then $d_S \leq d$, where d denotes the $GL(n,\mathbf{R})$ invariant metric on $P(n,\mathbf{R})$. Indeed $d_S(p,q) = |\log p - \log q|$, for any $p, q \in P(n,\mathbf{R})$ where $|Y|$ is defined on the linear space $S(n,\mathbf{R})$ by the Euclidean inner product $\operatorname{Tr} UV$. Inasmuch as one dimensional subspaces of $S(n,\mathbf{R})$ are geodesics in the Euclidean metric, and since $d_S = d$ along the one parameter subgroup $\exp tY$, we conclude that the one parameter subgroup p^t is the unique geodesic joining the identity element to p. Therefore between any two points there passes a unique geodesic. Moreover, the angles at the identity matrix with respect to both d_S and d coincide.

LEMMA 3.3. *If* a, b, c *are the sides of a geodesic triangle, then*

$$c^2 \geq a^2 + b^2 - 2ab\cos\measuredangle C\ ,$$

with equality if C *is at the identity, if and only if the triangle lies in the orbit of an abelian subgroup of* $P(n,\mathbf{R})$.

Proof. Put the vertex C at the identity via an isometry from $GL(n, \mathbb{R})$. By Lemma 3.2, the d_S distance for the opposite side does not exceed the d distance. The inequality now follows from the law of cosines for Euclidean space. The equality condition results from Lemma 3.2.

LEMMA 3.4. (i) *Let* A, B, C *be the vertices of a geodesic triangle and* a, b, c *the lengths of the corresponding sides. Then*

$$\sphericalangle A + \sphericalangle B + \sphericalangle C \leq 180^\circ$$

with equality if a vertex is at the identity, if and only if the triangle lies in the orbit of an abelian subgroup of $P(n, \mathbb{R})$.

(ii)
$$a \geq c \sin A\ .$$

If $\sphericalangle C \geq 90^\circ$, *then*

$$b \leq c \cos A\ .$$

Proof. (i) Consider a Euclidean triangle $A_1\ B_1\ C_1$ with sides of length a, b, c. By Lemma 3.3

$$\cos \sphericalangle C_1 = \frac{a^2 + b^2 - c^2}{2\,ab} \leq \cos \sphericalangle C\ .$$

Therefore $\sphericalangle C_1 \geq \sphericalangle C$. Similarly $\sphericalangle A_1 \geq \sphericalangle A$ and $\sphericalangle B_1 \geq \sphericalangle B$. Hence $\sphericalangle A + \sphericalangle B + \sphericalangle C \leq \sphericalangle A_1 + \sphericalangle B_1 + \sphericalangle C_1 \leq 180^\circ$. The equality condition follows from Lemma 3.3.

(ii) The first assertion comes from placing the vertex A at the identity and comparing the Riemannian length of the sides BC with the d_S-length. The second assertion of (ii) follows from

$$c^2 \cos^2 A = c^2 - c^2 \sin^2 A \geq a^2 + b^2 - c^2 \sin^2 A \geq b^2\ .$$

A subset of $P(n, \mathbb{R})$ is called a *geodesic subspace* if it contains for every pair of distinct points p_1 and p_2 the unique geodesic line passing through p_1 and p_2 (with respect to the $GL(n, \mathbb{R})$ invariant metric). For any pair of points p_1, p_2 we denote by $[p_1, p_2]$ the unique closed geodesic line segment from p_1 to p_2. A subset C of $P(n, R)$ is called *convex* if for every pair of points p_1, p_2 in C, the segment $[p_1, p_2]$ lies in C. The topological closure in $P(n, R)$ of a convex set is clearly convex.

We consider now the imbedding of $P(n, R)$ of the space $X = G/K$ where G is a semi-simple analytic linear group and K a maximal compact subgroup:

$$\mu(x\ K) = \rho(x)\ {}^t\rho(x)$$

where $\rho(G) = {}^t\rho(G)$. *The image* $\mu(X)$ *is a geodesic subspace.* For given $p_i \in \mu(X)$, we have $p_i = g_i\ {}^tg_i$ with $g_i \in \rho(G)$, $i = 1, 2$. The unique geodesic segment from p_1 to p_2 has the form $\{g_1 \exp s\ Y\ {}^tg_1;\ 0 \leq s \leq 1\}$ where $Y \in S(n, R)$. Hence $\exp Y = (g_1^{-1}\ g_2)\ {}^t(g_1^{-1}\ g_2) \in P(n, R) \cap \rho(G)$. By (2.3), $\exp s\ Y \in \rho(G)$ for all $s \in R$. Hence

$$g_1 \exp s\ Y\ {}^tg_1 = (g_1 \exp s/2\ Y)\ {}^t(g_1 \exp s/2\ Y) \in \mu(X)$$

for all s, and $\mu(X)$ contains the geodesic line passing through p_1 and p_2. More generally,

(3.4.1) *if* G *is an analytic group such that*

$$G = (G \cap P(n,R)) \cdot (G \cap O(n,R))$$

then $G \cap P(n,R)$ *is a geodesic subspace of* $P(n,R)$.

For any subset S of a metric space and for any non-negative real number v, we denote by $T_v(S)$ the subset of points lying within a distance less than v of S.

LEMMA 3.5. *Let* C *be a convex subset of* $P(n, R)$. *Then* $T_v(C)$ *is convex.*

Proof. Since $T_v(C) = T_v(\overline{C})$, we can assume without loss of generality that C is closed. Let $p_i \in T_v(C)$ $(i = 1,2)$. Since $C \cap T_v(p_i)$ is compact, there exists a point $q_i \in C$ with $d(p_i, C) = d(p_i, q_i)$ $(i = 1, 2)$.

Choose a point $p_0 \in [p_1, p_2]$ such that

$$d(p_0, [q_1, q_2]) = \sup_p d(p, [q_1, q_2]), \{p \in [p_1, p_2]\} ,$$

with p_0 taken as an endpoint if the supremum is attained at either p_1 or p_2. Choose a point $q_0 \in [q_1, q_2]$ such that $d(p_0, q_0) = d(p_0, [q_1, q_2])$. We consider three cases

Case 1. p_0 *is an endpoint.* Here

$$v > d(p_0, [q_1, q_2]) \geq d(p, [q_1, q_2])$$

for all $p \in [p_1, p_2]$ so that $[p_1, p_2] \subset T_v([q_1, q_2])$.

Case 2. p_0 *is not an endpoint and* q_0 *is an endpoint of* $[q_1, q_2]$. For definiteness, suppose $q_0 = q_1$. Since $d(p_0, q_1) \geq d(p_1, q_1)$, the angle at p_0 in the triangle $p_1\ p_0\ q_1$ cannot be obtuse by Lemma 3.3. Hence $\measuredangle p_2 p_0 q_0 \geq 90°$.

If $q_1 = q_2$, then $d(p_2, q_1) \geq d(p_0, q_1)$ by Lemma 3.3 and thus $d(p_2, [q_1, q_2]) = d(p_0, [q_1, q_2])$. This is impossible, by choice of p_0. Thus we can suppose $q_1 \neq q_2$.

We claim $\measuredangle p_0 q_0 q_2 \geq 90°$. Otherwise by selecting a point $q \in [q_1, q_2]$ near q_0, we would get an obtuse angle for $\measuredangle p_0 q\, q_0$ and therefore $d(p_0, q) < d(p_0, q_0)$ by Lemma 3.3 – contradicting $d(p_0, q_0) = d(p_0, [q_1, q_2])$.

We now consider the quadrilateral $p_2 p_0 q_0 q_2$ with angles at p_0 and q_0 at least 90°. Set $c = d(q_0, p_2)$, $\theta = \measuredangle p_2 q_0 q_2$, $\theta' = \measuredangle p_2 q_0 p_0$. Applying Lemma 3.4 (ii), and noting that we get $\theta + \theta' \geq \measuredangle p_0 q_0 q_2$,

$$d(p_0, q_0) \leq c \cos \theta' = c \sin (90 - \theta') \leq c \sin \theta \leq d(p_2, q_2) .$$

Thus we could have selected p_2 as p_0 – a contradiction. Hence Case 2 does not occur.

Case 3. p_0 *is not an endpoint of* $[p_1, p_2]$ *and* q_0 *is not an endpoint of* $[q_1, q_2]$.

Since $d(p_0, q_0) = d(p_0, [q_1, q_2])$, we have $\measuredangle p_0 q_0 q_i \geq 90^\circ$ for $i = 1, 2$, by the argument above based on Lemma 3.3.

Also $\measuredangle p_1 p_0 q_0 + \measuredangle p_2 p_0 q_0 = 180^\circ$, so that one of these two angles is at least 90°. Assume for definiteness that $\measuredangle p_2 p_0 q_0 \geq 90^\circ$. Then the quadrilateral $p_2 p_0 q_0 q_2$ has angles at p_0 and q_0 at least 90°. Therefore, as above $d(p_0, q_0) \leq d(p_2, q_2)$ and we get the same contradiction. Thus Case 3 does not occur.

Since only Case 1 occurs, Lemma 3.5 is proved.

REMARK. Let F be a geodesic subspace of the space $P(n, \mathbf{R})$. For any point $p \in P(n, \mathbf{R})$ we denote by $\pi(p)$ a point in F (which is closed) such that

$$d(p, \pi(p)) = d(p, F) .$$

Then the geodesic segment $[p, \pi(p)]$ forms a right angle at $\pi(p)$ with all geodesics in F through $\pi(p)$, by Lemma 3.3. It follows from the fact that the sum of the angles in a geodesic triangle is at most 180° that the point $\pi(p)$ is unique. We call the map $\pi: P(n, \mathbf{R}) \to F$ the *orthogonal projection* of $P(n, \mathbf{R})$ onto F. For any points p_1 and p_2 in $P(n, \mathbf{R})$ the quadrilateral $p_1\, \pi(p_1)\, \pi(p_2)\, p_2$ has right angles at $\pi(p_1)$ and $\pi(p_2)$; therefore $d(p_1, p_2) \geq d(\pi(p_1), \pi(p_2))$.

DEFINITION. Let L be a geodesic line in the Riemannian space $P(n, \mathbf{R})$ and let f be a real valued function on L. Let $s \to p(s)$ be a parametrization of L by arc length. The function f is called *convex* if for all a and b in $\mathbf{R}$ and t in $[0, 1]$,

$$f(p((1-t)\,a + t\,b)) \leq (1-t)\, f(p(a)) + t\, f(p(b)) \tag{3.6.1}$$

LEMMA 3.6. *Let* L *be a geodesic line and* C *a convex set in* $P(n, \mathbf{R})$. *Then* $p \to d(p, C)$ *is a convex function on* L.

Proof. We lose no generality in assuming that C is closed.

Given any point p in $P(n,\mathbf{R})$, there is a *unique* point q in C such that $d(p,q) = d(p,C)$. For suppose otherwise, then there would be $q_1 \in C$ and $q_2 \in C$ with $q_1 \neq q_2$ and $d(p,q_1) = d(p,q_2) = d(p,C)$. By Lemma 3.5, the closed ball of center p and radius $d(p,q_1)$ is convex and therefore contains the segment $[q_1,q_2]$. It follows therefore that $d(p,q) = d(p,[q_1,q_2])$ for all $q \in [q_1,q_2]$. Selecting q in the interior of $[q_1,q_2]$, the segment $[p,q]$ forms an angle of at least 90° with either $[q,q_1]$ or $[q,q_2]$ – say $[q,q_1]$ for definiteness. Then $d(p,q_1) > d(p,q)$ by Lemma 3.3. This contradiction implies that q is unique.

Given $p_i \in L$, select $q_i \in C$ so that $d(p_i,C) = d(p_i,q_i)$ $(i = 1,2)$. It suffices to prove Lemma 3.6 for the case $C = [q_1,q_2]$. For $d(p,C) \leq d(p,[q_1,q_2])$ with equality at $p = p_1$ and $p = p_2$, so that one condition (3.6.1) is verified for $d(p,[q_1,q_2])$ on the segment $[p_1,p_2]$, it follows for $d(p,C)$. Thus we assume that $C = [q_1,q_2]$.

For each point p in $P(n,\mathbf{R})$, let $q(p)$ denote the nearest point in $[q_1,q_2]$ to p. It is easy to verify that $p \to q(p)$ is a differentiable function on the complement of $[q_1,q_2]$ (namely, consider it separately on the three connected components of the complement of $\pi^{-1}(q_1) \cup \pi^{-1}(q_2)$, where π is the orthogonal projection onto the geodesic line containing $[q_1,q_2]$).

Let $s \to p(s)$ be a parametrization of the line L with $p(a) = p_1$ and $p(b) = p_2$, and set $f(s) = d(p(s),[q_1,q_2])$. Then $f(s) = d(p(s), q(p(s)))$ and if $f(s_0) \neq 0$, we have $f'(s_0) = f_1(s_0) + f_2(s_0)$ where $f_1(s_0) = \frac{d}{ds} d(p(s), q(p(s_0)))|_{s_0}$ and $f_2(s_0) = \frac{d}{ds} d(p(s_0), q(p(s)))|_{s_0}$. From the definition of $q(p)$, $d(p(s_0), q(p(s_0))) \leq d(p(s_0), q(p(s)))$ for all s, and hence $f_2(s_0) = 0$ if $f(s_0) \neq 0$. Thus if $f(s) \neq 0$,

$$f'(s) = f_1(s) = \sin(\theta(s) - 90) = -\cos\theta(s) \tag{3.6.2}$$

where $\theta(s)$ is the angle $q(p(s))\,p(s)\,p_2$. Therefore

$$f''(s) = \sin\theta(s) \cdot \frac{d\theta}{ds}\ .$$

Consider now the quadrilateral $q(p(s_0))\, p(s_0)\, p(s_0+\Delta s)\, q(p(s_0+\Delta s))$. Its angles at $q(p(s_0))$ and $q(p(s_0+\Delta s))$ are at least $90°$, otherwise by Lemma 3.3 one could find nearer points to $p(s_0)$ and $p(s_0+\Delta s)$ on $[q_1, q_2]$ than $q(s_0)$ and $q(s_0+\Delta s)$. Since the sum of the four angles of the quadrilateral does not exceed $360°$, it follows at once that $\theta(s_0+\Delta s) \geq \theta(s_0)$ and thus $f''(s) \geq 0$ wherever $f(s) \neq 0$.

Case 1. $L \cap [q_1, q_2]$ *is empty*. Here $f''(s) \geq 0$ for all s and thus f is a convex function.

Case 2. $L \cap [q_1, q_2]$ *is not empty*. Then if it consists of a single point, f vanishes at a single value s_0, and f is convex on $[-\infty, s_0]$ and $[s_0, \infty]$ with $f(s_0) = 0$. Since $f(s) \geq 0$ for all s, f is monotonically decreasing on $[-\infty, s_0]$, monotonically increasing on $[s_0, \infty]$ and convex on $\mathbf{R}$. If $L \cap [q_1, q_2]$ contains more than one point, then $[q_1, q_2] \subset L$ and the convexity of f is clear.

The proof of Lemma 3.6 is now complete.

LEMMA 3.7. *Let* L *be a geodesic line and* F *a geodesic subspace such that* $L \subset T_v(F)$ *for some finite* v. *Then* $d(p, F) = d(L, F)$ *for all* $p \in L$. *Moreover, for any distinct points* p_1 *and* p_2 *in* L, *the four angles of quadrilateral* $p_1\, \pi(p_1)\, \pi(p_2)\, p_2$ *are right angles, where* π *is the orthogonal projection onto* F.

Proof. Let $f(s) = d(p(s), F)$ as in the proof of Lemma 3.6. Then f is a bounded convex function on $\mathbf{R}$. It follows at once that f is constant. From this the first assertion of the Lemma follows. The second assertion follows by Lemma 3.3 from the fact that $d(p_i, \pi(p_i)) = d(p_i, F) = d(L, \pi(p_i))$, $i = 1, 2$; or alternatively, it follows from (3.6.2).

REMARK. Inasmuch as Lemmas 3.3 through 3.7 are valid for any geodesic subspace of $P(n, \mathbf{R})$, they are valid for the symmetric Riemannian space X associated to a semi-simple analytic group.

DEFINITION. A geodesic subspace F of $P(n, \mathbf{R})$ is called *flat* if and only if the sum of the three angles of every geodesic triangle in F is $180°$.

Let ABC be a non-degenerate geodesic triangle whose angles have sum $180°$. *Then the triangle lies in a flat subspace.* To see this, we lose no generality in assuming that A is the identity matrix, for we can move A into the identity by an isometry. The sides AB and AC are then given by $\exp sY$ and $\exp tZ$ with Y and Z in $S(n, \mathbf{R})$. By Lemma 3.3, Y and Z commute. It follows at once that the subset $\{\exp sY + tZ, s, t \in \mathbf{R}\}$ is isometric to $\mathbf{R}^2$. Hence it is a flat subspace of $P(n, \mathbf{R})$ containing A, B, C; we call such a subspace a *flat 2-plane.*

Let now $ABCD$ be a quadrilateral the sum of whose angles is $360°$. Upon drawing the diagonal AC, we get two geodesic triangles each of whose angle sums is not less than $180°$. It follows that each angle sum is $180°$ by Lemma 3.4 (i), and that moreover

$$\measuredangle BAC + \measuredangle CAD = \measuredangle BAD \ .$$

Therefore ABC and ACD lie in flat two dimensional geodesic subspaces. Furthermore upon drawing the diagonal BD, we can infer that BAD lies in a flat two-dimensional plane. From $\measuredangle BAC + \measuredangle CAD = \measuredangle BAD$, we conclude that the three 2-planes coincide. Thus $ABCD$ lies in a flat 2-plane.

§4. A Metric Definition of the Maximal Boundary

Let G be a semi-simple analytic group having a faithful matrix representation, let K be a maximal compact subgroup, and set $X = G/K$. Let x_K denote the point of X fixed by K. As pointed out in 2.11, we can find a faithful representation ρ of G into $GL(n, R)$ so that $\rho(G) = {}^t\rho(G)$ and $\rho(K) = \rho(G) \cap O(n, R)$. The map $\mu : xK \to \rho(x)\, {}^t\rho(x)$ then yields an imbedding of G onto a geodesic subspace of $P(n, R)$ and the induced metric on X is G-invariant. It is convenient sometimes to identify X with $\mu(X)$.

A geodesic subspace F of X is called *flat* if and only if the sum of the three angles of every geodesic triangle in F is $180°$.

REMARK. Although X admits more than one G-invariant metric (cf. Remark 1 of Section 3), the notion of *flat* is the same for all G-invariant metrics.

The flat subspaces through x_K are precisely the orbits of x_K under abelian analytic subgroups contained in $\rho^{-1}(P(n, R) \cap \rho(G))$; the flat subspaces through gx_K are therefore the orbits of gx_K under the polar subgroups gAg^{-1} where A is a polar subgroup contained in $\rho^{-1}(P(n, R) \cap \rho(G))$. Inasmuch as the maximal polar subgroups of $\rho(G)$ are conjugate under an inner automorphism, we conclude that G permutes transitively all the maximal flat subspaces of X. From 2.6 (v) we conclude that K permutes transitively all the maximal flat subspaces of X passing through the point x_K.

DEFINITION. The *rank* of X is the dimension of a maximal flat subspace.

Since $Ax_K = AK/K = A/A \cap K = A$ for any polar subgroup of $\rho(G)$, we see that

$$\text{rank } X = \text{TR-rank } G .$$

Moreover any polar subgroup operates freely in X. Indeed $p_1 pK = p_2 pK$ with $\rho(p_i) \in P(n, R) \cap \rho(G)$ implies $p p_1^2 p = p p_2^2 p$ and hence $p_1 = p_2$.

If rank $X = r$, we call a maximal flat geodesic subspace an *r-flat*. A geodesic line is called *regular* if it lies in only one r-flat; it is called *singular* if it is not regular. Since the orbit of 1 under a maximal polar subgroup of $\rho(G) \cap P(n, R)$ is an r-flat of $\mu(X)$, we see that a geodesic line through 1 is singular in $\mu(X)$ if and only if it is the orbit of an TR-singular polar subgroup of $\rho(G) \cap P(n, R)$.

Assume now that ρ = identity; identify X with $\mu(X)$.

Let F be an r-flat through x_K. Let A be the maximal polar subgroup of $G \cap P(n, R)$ such that $Ax_K = F$. Then the union of singular geodesics in F through x_K is the set Sx_K where S is the set of R-singular elements in A. Let $^{\blacktriangleleft}F$ be a topologically connected component of $F - Sx_K$. Then $^{\blacktriangleleft}F = {}^{\blacktriangleleft}Ax_K$ where $^{\blacktriangleleft}A$ is a chamber in A.

DEFINITION. Let F be an r-flat in X. Let $x \in X$. Let ^xS denote the union of singular geodesic rays through x. A topologically connected component of $F - {}^xS$ is called a *chamber* of origin x. The topological closure of a chamber is called a *closed chamber*.

From the foregoing, it is clear that every chamber of origin gx_K has the form $g^{\blacktriangleleft}Ax_K$ where $^{\blacktriangleleft}A$ is a chamber in $G \cap P(n, R)$. By (2.6) (v) and the last sentence of (2.2), all the chambers in $G \cup P(n, R)$ are conjugate under K. Therefore G operates transitively on the set of all chambers in X.

DEFINITION. A *chamber wall* of a chamber $g^{\blacktriangleleft}Ax_K$ is a subset $g^{\blacktriangleleft}Bx_K$, where $^{\blacktriangleleft}B$ is a wall of the chamber $^{\blacktriangleleft}A$ in G.

NOTATION. For any subsets A and B in X we denote by hd(A, B) the *Hausdorff distance* between A and B; i.e.,

$$hd(A, B) = \inf \{v \leq \infty; A \subset T_v(B), B \subset T_v(A)\} .$$

LEMMA 4.1. *Let* X_0 *denote the set of all chambers in* X. *Define two chambers* $\blacktriangleleft F_1$ and $\blacktriangleleft F_2$ *to be equivalent if and only if*

$$hd(\blacktriangleleft F_1, \blacktriangleleft F_2) < \infty .$$

Define X_0 *to be the quotient of* X_0 *by this equivalence relation. Then* $X_0 = G/P$ *where* P *is a minimal parabolic subgroup of* G.

Proof. Let K be a maximal compact subgroup of G. We may assume that $G = G \cap P(n,\mathbf{R}) \cdot K$ with $K = G \cap O(n,\mathbf{R})$. Let A be a maximal polar subgroup of $G \cap P(n,\mathbf{R})$. Let $\blacktriangleleft A$ be a chamber in A. Set $U = U(\blacktriangleleft A)$ and $P = P(\blacktriangleleft A)$ (cf. (2.4)). Then $P = Z(A)U$ and P is a minimal parabolic subgroup. Furthermore, $G = KP$ by (2.7).

From the definition of $U(\blacktriangleleft A)$ it is clear that for any $u \in U$, $\{a^{-1}ua; a \in \blacktriangleleft A\}$ is a bounded (i.e., relatively compact) set in U. Therefore for any $p \in P$, $\{a^{-1}pa; a \in \blacktriangleleft A\}$ is bounded. Set $\blacktriangleleft F = \blacktriangleleft A x_K$. Then for

$$\begin{aligned} p \in p,\ hd(p\blacktriangleleft F, \blacktriangleleft F) &\leq \sup_a d(pax_K, ax_K)\,\{a \in \blacktriangleleft A\} \\ &\leq \sup_a d(a^{-1}pax_K, x_K) \\ &< \infty . \end{aligned}$$

For any $k \in K$ such that $k\blacktriangleleft F \neq \blacktriangleleft F$, we can find a geodesic ray L in $k\blacktriangleleft F$ that is not in $\blacktriangleleft F$. Then in the Euclidean metric d_S defined by $\mathrm{Tr}\,(\log p)^2$, we find $d_S(L - T_r(1), \blacktriangleleft F) \geq cr$ for some constant c. Therefore $hd(\blacktriangleleft F, k\blacktriangleleft F) = \infty$. The condition $k\blacktriangleleft F = \blacktriangleleft F$ is equivalent to $k\blacktriangleleft AK = \blacktriangleleft AK$ or $k\blacktriangleleft Ak^{-1} \cdot K = \blacktriangleleft A \cdot K$ or, by (2.6) (i), $k\blacktriangleleft Ak^{-1} = \blacktriangleleft A$. This last condition is equivalent to $k \in Z(A)$ by (2.3) (iv). Thus if $k \in K$, $hd(k\blacktriangleleft F, \blacktriangleleft F) < \infty$ if and only if $k \in K \cap P = Z(A) \cap K$.

Given now an element g in G but not in P, we have $g = pk$ with $p \in P$ and $k \in K - P$ since $g = G^{-1} = (KP)^{-1} = PK$. Therefore

$$\begin{aligned} hd(g\blacktriangleleft F, \blacktriangleleft F) &= hd(pk\blacktriangleleft F, \blacktriangleleft F) \\ &\geq hd(pk\blacktriangleleft F, p\blacktriangleleft F) - hd(p\blacktriangleleft F, \blacktriangleleft F) \\ &\geq hd(k\blacktriangleleft F, \blacktriangleleft F) - hd(p\blacktriangleleft F, \blacktriangleleft F) \\ &= \infty . \end{aligned}$$

Let $[\blacktriangleleft F]$ denote the equivalence class of $\blacktriangleleft F$ in X_0. We have proved that P is the stabilizer of $[\blacktriangleleft F_0]$. This proves Lemma 4.1.

We define the *topology* of X_0 to be the quotient space topology of G/P.

LEMMA 4.2. *Let* $\blacktriangleleft F_0$ *and* $\blacktriangleleft F$ *be chambers or chamber walls in* X *of origins* p_0 *and* p *respectively such that* $hd(\blacktriangleleft F_0, \blacktriangleleft F) < \infty$. *Then*

$$hd(\blacktriangleleft F_0, \blacktriangleleft F) \leq d(p_0, p) \ .$$

Proof. Let L be a half-line in $\blacktriangleleft F_0$ of origin p_0. Then the function $q \to d(q, \blacktriangleleft F)$ is convex on L by Lemma 3.6; since it is also bounded, it must be monotonically decreasing and therefore takes on its maximum value at p_0. Inasmuch as $\blacktriangleleft F_0$ is the union of half-lines of origin p_0, we get

$$d(p_0, \blacktriangleleft F) \geq d(q, \blacktriangleleft F) \tag{4.2.1}$$

for all $q \in \blacktriangleleft F_0$. Similarly $d(p, \blacktriangleleft F_0) \geq d(q, \blacktriangleleft F_0)$ for all $q \in \blacktriangleleft F$. Hence

$$\begin{aligned} hd(\blacktriangleleft F_0, \blacktriangleleft F) &= \sup\{d(p_0, \blacktriangleleft F), d(\blacktriangleleft F_0, p)\} \\ &\leq d(p_0, p) \end{aligned}$$

by applying the above convexity argument to the function $q \to d(p_0 q)$ restricted to $\blacktriangleleft F$.

§5. Polar Parts

We continue the notation G, K, and X *and the assumptions made in* Section 4. Let F be a flat subspace of X. Let $\blacktriangleleft F$ be a chamber or chamber wall in F, let E be a geodesic subspace. Set

$$G_E = \{g \in G;\ gE \subset E\}$$
$$G_{\blacktriangleleft F} = \{g \in G;\ g\blacktriangleleft F \subset \blacktriangleleft F\}\ .$$

One sees easily that $gE = E$ for all $g \in G_E$ and $G_{\blacktriangleleft F} \subset G_F$. Thus $G_{\blacktriangleleft F}$ is a semi-group and G_E is a subgroup.

LEMMA 5.1. *Let* $r = \text{rank}\ X$. *Let* F *be a flat subspace of* X. *Then* G_F *has a unique maximal polar subgroup, denoted* pol F, *which contains the polar part of every element of* G_F *and which operates simply transitively on* F. *Moreover, if* F *is an* r*-flat, then* $G_F = N(\text{pol}\ F)$. *The map* $F \to \text{pol}\ F$ *of the set of all* r*-flats to the set of all maximal polar subgroups is bijective.*

Proof. Let x_K be the point of X stabilized by $K = G \cap O(n, R)$. Since G is transitive on X, no generality is lost in assuming $x_K \in F$. Then $F = Bx_K$ where B is a polar subgroup containing in $G \cap P(n, R)$ by Lemma 3.4 (i). Since B is transitive on F, we get $G_F = (G_F \cap K)B$ since $B \cap K = (1)$, we see that B is simply transitive on F. For any $k \in G_F \cap K$, we have $kBK \subset BK$. Therefore, $kBk^{-1} \subset BK$. From this we infer $kBk^{-1} = B$ by (2.6) (i). Thus B is normal in G_F. Inasmuch as the maximal polar subgroups of G_F are conjugate via an inner automorphism of G_F, it follows that B is the unique maximal polar subgroup of G_F. Clearly $N(B) \cap K \subset G_F$. Thus $G_F = (N(B) \cap K)B$.

The flat subspace F is an r-flat if and only if B is a maximal polar subgroup of G. In that case, $N(B) = (N(B) \cap K)B$ by (2.6). Thus $G_F = N(pol\, F)$ if F is an r-flat.

Suppose now that F and F_1 are r-flats stabilized by *pol* F. Since G operates transitively on the r-flats, we have $F_1 = gF$ for some g in G. Hence $G_{F_1} = G_{gF} = g G_F g^{-1}$. Since *pol* F_1 is the set of all polar elements in G_{F_1}, we see that *pol* $F_1 = g$ *pol* F g^{-1}. By hypothesis *pol* $F \subset G_{F_1}$. Since *pol* F is a maximal polar subgroup, we get *pol* F = *pol* F_1 and therefore *pol* F = g *pol* F g^{-1}; that is, $g \in N(pol\, F)$. Since $G_F = N(pol\, F)$, we get $g \in G_F$ and $F_1 = g F = F$. Thus $F \to pol\, F$ is a bijective map.

REMARK. Let p and q be distinct points of X and let σ_p and σ_q be the corresponding symmetries of X. Let L be the geodesic line through p and q. It is easy to verify that $\sigma_q \sigma_p$ is the element in *pol* L which translates p to a point q_1 such that q is the midpoint of $[p, q_1]$. Namely, if $q = gp$, then $\sigma_q = g\, \sigma_p\, g^{-1}$ and $\sigma_q \sigma_p = g(\sigma_p\, g^{-1}\, \sigma_p)$ without loss of generality, we can assume that $p = x_K$ and $\sigma_p(g) = {}^t g^{-1}$. Then $\sigma_q \sigma_p = g\, {}^t g \in G \cap P(n, \mathbf{R})$, and thus $g\, {}^t g \in pol\, L$ where L is the line through p and q.

DEFINITION. Let S_1 and S_2 be flat subspaces in X. We say that S_2 is a *parellel translate* of S_1 if and only if S_1 and S_2 lie in a flat subspace and $S_2 = pS_1$ with $p \in pol\, F$ for any flat space F containing S_1 and S_2.

REMARK. Let S be a flat subspace of X and let $g \in Z(pol\, S)$. Then gS *is a parallel translate of* S.

For let $p = pol\, g$, and let B be the polar subgroup generated by *pol* S and $\{p^t;\ t \in \mathbf{R}\}$. No generality is lost in assuming $B \subset G \cap P(n, \mathbf{R})$. In this case, Bx_K is a flat subspace containing pS and S.

Set $G_1 = Z(pol\, S)$.

Let $g = kp$. Then $pk = kp$ and ${}^tkp = p^tk$. From this we deduce that p commutes with $(k^tk)^{\frac{1}{2}}$. Thus writing $k = q \cdot c$ with $q \in G_1 \cap P(n, R)$ and $c \in G_1 \cap O(n, R)$, we see that p commutes with q and c. We have $cS = c\ pol\ Sx_K = pol\ Scx_K = pol\ Sx_K = S$ and thus $gS = pkS = pqS$. The polar subgroups $pol\ S$, $\{p^t; t \in R\}$, and $\{q^t; t \in R\}$ commute elementwise; let C denote the polar subgroup of $G_1 \cap P(n, R)$ which they generate. Set $E = Cx_K$. Then $pq \in C = pol\ E$ and thus pqS is a parallel translate of S in the usual Euclidean sense in the flat space E. This implies there is an element $b \in B$ such that $bS = gS$. Now for any flat subspace F containing S and gS, $pol\ F$ contains B. From this our remark follows.

LEMMA 5.2. *Let* G *be a semi-simple analytic linear group,* K *a maximal compact subgroup, and* $X = G/K$. *Assume* $\operatorname{rank} X = r$. *Then*

(i) *A polar regular element of* G *is semi-simple.*

(ii) *A polar regular element of* G *stabilizes a unique* r*-flat in* X.

(iii) *Let* g *be a polar regular element of* G *stabilizing the* r*-flat* F *in* X. *Then* $Z(g)$ *stabilizes and acts transitively on* F. *Moreover,* $gx = (pol\ g)x$ *for all* $x \in F$.

Proof. Let g be a polar regular element of G and let A be the unique maximal polar subgroup containing $pol\ g$. Since the maximal polar subgroups are conjugate, we lose no generality in assuming $A \subset G \cap P(n, R)$ with $K = G \cap O(n, R)$.

Set $s = pol\ g$. Then $Z(s) = {}^tZ(s)$ and therefore $Z(s) = Z(s) \cap P(n, R) \cdot Z(s) \cap O(n, R)$ by (2.6). Since s lies in a *unique* maximal polar subgroup of G, we conclude that $Z(s) \cap P(n, R) = A$ and that $Z(s) \subset N(A)$. Let F be an r-flat such that $pol\ F = A$. Then $N(A) = G_F$ by Lemma 5.1, and hence $g \in G_F$; that is, g stabilizes an r-flat if and only if A does. If A stabilizes an r-flat F', then $A = pol\ F'$ by Lemma 5.1 and thus $G_F = N(A) = G_{F'}$. Let h be an element in G with $hF = F'$. Then

$h \underset{F}{G} h^{-1} = G_{hF} = G_{F'} = G_F$; hence $h \in N(A) = G_F$ and $F' = hF = F$. This proves (ii). Assertion (i) follows from a sharpening of the observation above that $Z(s) \subset N(Z)$. Namely, the group $Z(s)$ operating by inner-automorphisms, stabilizes not only A, but the *chamber* containing s. By (2.3) (iv), $N(A)/Z(A)$ operates *simply* transitively on the chambers of A. It follows at once that $Z(s) = Z(A)$. We have

$$Z(A) = A \cdot (Z(A) \cap K) .$$

This decomposition corresponds to the polar decomposition of the elements of $Z(A)$, and we see therefore that $Z(A)$ has no unipotent elements other than (1). Since $g \in Z(s)$, we see that g is semi-simple.

Set $k = s^{-1}g$. Then $Z(g) = Z(s) \cap Z(k)$ and is a subgroup of $Z(A)$ containing A. This implies (iii).

LEMMA 5.3. *Let* G *be a closed self-adjoint subgroup of finite index in an algebraic subgroup of* $GL(n, \mathbf{R})$. *Set* $K = G \cap O(n, \mathbf{R})$ *and let* $X = G/K$. *Let* $\mu : X \to P(n, \mathbf{R})$ *be the injection given by* $\mu(gK) = g^t g$. *Let* d *denote the metric induced from the metric* $ds^2 = \mathrm{Tr}\,(p^{-1}\dot{p})^2$ *via* μ. *Then for any element* $g \in G$,

$$\inf_{x \in X} d(x, gx)^2 = 4\, \mathrm{Tr}\,(\log pol\ g)^2 .$$

Moreover, the infinimum is attained if and only if g *is semi-simple.*

Proof. Set $|g|^2 = \mathrm{Tr}\,(\log pol\ g)^2 = \Sigma_\lambda \log |\lambda|^2$ where λ ranges over the n eigenvalues of g. The proof of Lemma 5.3 will come after a few observations.

(5.3.1) *For any element* g *and* h *in* G,

$$\inf_{x \in X} d(x, gx) = \inf_{x \in X} d(hx, ghx) = \inf_{x \in X} d(x, h^{-1}ghx)$$

and

$$|g| = |h^{-1}gh| .$$

(5.3.2) *Let* Y *be a geodesic subspace of* X *stable under* g. *Then*

$$\inf_{x \in X} d(x, gx) = \inf_{y \in Y} d(y, gy) .$$

For given $x \in X$, let y denote the foot of the perpendicular from x to Y. The geodesic quadrilateral y, gy, gx, x has right angles at y and gy. Set $c = d(y, gx)$, and set $\theta = \measuredangle(gx)y(gy)$, $\theta' = \measuredangle xy(gx)$. By Lemma 3.4, $d(x, gx) \geq c \sin \theta'$ and $d(y, gy) \leq c \cos \theta$. Since $\theta' + \theta = 90°$, we get $d(x, gx) \geq d(y, gy)$. This proves (5.3.2).

Let $g \in G$. Then $g = k \cdot p \cdot u$ where $p = \mathit{pol}\, g$, u is the unipotent Jordan component, and k, p, u commute. Let $s = kp$. Then s is the semi-simple Jordan component. Since p is conjugate by an inner automorphism of G to an element in $G \cap P(n, \mathbf{R})$, no generality is lost in assuming $p \in G \cap P(n, \mathbf{R})$. Since k lies in a compact subgroup of $G \cap Z(p)$, k is conjugate by an inner automorphism of the self-adjoint group $G \cap Z(p)$ to an element in $O(n, \mathbf{R}) \cap G \cap Z(p)$. Thus without loss of generality we can assume $k \in Z(p) \cap O(n, \mathbf{R})$, by (5.3.1). Thus $Z(s) = Z(k) \cap Z(p) = {}^tZ(s)$ and $u \in Z(s) \cap G$. Let $Z = Z(s) \cap G$. The group Z decomposes into $(Z \cap P(n, \mathbf{R}))(Z \cap O(n, \mathbf{R})$ and therefore the orbit of the point K under Z is a geodesic subspace of X. By (5.3.2), we have $\inf_{x \in X} d(x, gx) = \inf_{z \in Z} d(zK, gzK)$. Since $Z \subset Z(k)$, we have

$$\inf_{z \in Z} d(zK, gzK) = \inf_{z \in Z} d(zK, kpuzK) = \inf_{z \in Z} d(zK, puzK) .$$

By the Jacobson-Morozov theorem (cf. (2.9)) the element u can be embedded in an analytic subgroup S of Z locally isomorphic to $SL(2, \mathbf{R})$, and accordingly there is a sequence of elements $s_n \in S$ such that $s_n^{-1} u\, s_n$ converges to the identity element. Hence

$$\inf_{z \in Z} d(zK, puzK) \leq \inf_n d(s_nK, pus_nK) =$$

$$\inf_n d(K, ps_n^{-1}us_nK) = d(K, pK) .$$

On the other hand, let B denote the one parameter subgroup $\{p^t; t \in \mathbf{R}\}$ of $P(n, \mathbf{R}) \cap G$. Then the orbit of the point K under B is a geodesic in X by the remark following Lemma 3.2. By (5.3.2) we have

$$\inf_{x \in X} d(x, px) = \inf_{b \in B} d(bK, pbK) = d(K, pK) .$$

It follows at once that

$$\inf_{x \in X} d(x, gx) \leq d(K, pK) .$$

In case g is semi-simple we have $u = 1$ and $gBK = kpBK = BK$. Thus BK is a flat geodesic subspace stable under g; and for all $y \in BK$ we have $gy = py$ and $d(gy, y) = d(py, y) = d(pK, K) = |g|$. This proves

(5.3.3). *A semi-simple element keeps invariant a non-empty flat geodesic subspace. For any minimal flat geodesic* g*-stable subspace* F *and for any* $y \in F$,

$$d(gy, y) = \inf_{x \in X} d(gx, x) .$$

Moreover, if $d(gy, y) = \inf_{x \in X} d(gx, x) \neq 0$, *then the geodesic line through* y *and* gy *is stable under* g *and* pol g.

(5.3.4). *Let* $p \in P(n, \mathbf{R})$ *and let* v *be a unipotent element in* $GL(n, \mathbf{R})$ *such that* $pv = vp$. *Then*

$$\operatorname{Tr} (\log pv^t vp)^2 = \operatorname{Tr} (\log p^2)^2 + \operatorname{Tr} (\log v^t v)^2 .$$

Proof. For any $k \in O(n, \mathbf{R})$ and for any $g \in GL(n, \mathbf{R})$, ${}^t(kgk^{-1}) = k^tgk^{-1}$. Hence no generality is lost in replacing the canonical base of $\mathbf{R}^n$ by any other orthonormal base.

Let $\Sigma_\lambda V_\lambda$ be the decomposition of $\mathbf{R}^n$ into the eigenspaces of the distinct eigenvalues $\lambda_1, \ldots, \lambda_s$ of p. Then each V_λ is stable under v and ${}^tv$ since $pv = vp$ and ${}^tvp = p^tv$. Thus we have $p = \Sigma p_\lambda$ and $v = \Sigma v_\lambda$ as direct sum decompositions where p_λ operates as multiplication by λ on V_λ and v_λ is the restriction of v to V_λ. Inasmuch as V_λ

is orthogonal to V_λ, if $\lambda \neq \lambda'$, we can select an orthonormal base for R^n by combining orthonormal bases for each V_λ. Thereby, the proof of (5.3.4) is reduced to the verification for each pair p_λ and v_λ. Thus, without loss of generality, we may assume that $p = \lambda \cdot 1$.

Inasmuch as $\det v^t v = 1$, we find that $\operatorname{Tr} \log v^t v = 0$. Therefore

$$\begin{aligned} \operatorname{Tr} (\log pv^t vp)^2 &= \operatorname{Tr} (\log p^2 v^t v)^2 \\ &= \operatorname{Tr} (\log p^2 + \log v^t v)^2 \\ &= \operatorname{Tr} (2 \log \lambda)^2 + \operatorname{Tr} (\log v^t v)^2 \\ &= \operatorname{Tr} (\log p^2)^2 + \operatorname{Tr} (\log v^t v)^2 \ . \end{aligned}$$

Applying (5.3.4) to $d(zK, puzK) = d(K, pz^{-1}uzK)$ with $v = z^{-1}uz$, we obtain

$$d(zK, puzK)^2 = d(K, pK)^2 + d(K, vK)^2 \ .$$

Therefore

$$\inf_{x \in X} d(x, gx) = \inf_{z \in Z} d(zK, puzK) \geq d(K, pK) \ .$$

It now follows that $d(K, pK) = \inf_x d(x, gx), \{x \in X\}$ with the infimum attained if and only if g is semi-simple. Clearly $d(K, pK)^2 = \operatorname{Tr}(\log p^2)^2$. Lemma 5.3 is now proved.

Let E be a geodesic subspace of X. Then for any $p \in E$, the symmetry of X with respect to p stabilizes E, that is $\sigma_p \in G'_E$, where $G' = \sigma_p G \cup G$ (cf. 2.10). Inasmuch as G'_E is generated by $\{\sigma_p; p \in E\}$ (cf. Remark 2 of Section 3), we see that G_E operates transitively on E. Let *pol* E denote the subgroup generated by *pol* L for all geodesic lines L in E. It is not hard to see that *pol* E is the subgroup of G_E generated by all its polar subgroups. (cf. Remark following Lemma 5.1.)

In particular, *pol* $E \subset G_E$. Also, if $E \subset F$, then *pol* $E \subset$ *pol* F. Finally if G and G_E are self-adjoint groups, then the group *pol* E is a self-adjoint group and

$$(5.4.1)\qquad \begin{aligned} pol\,E &= (pol\,E \cap P(n,R)) \cdot (pol\,E \cap O(n,R)) \\ G_E &= (pol\,E \cap P(n,R)) \cdot (G_E \cap O(n,R)) \,. \end{aligned}$$

DEFINITION. Let E and F be geodesic subspaces in X. We call E *a parallel translate of* F if and only if $E = gF$ with $g \in Z(pol\,F)$.

This definition is an extension of the definition for flat subspaces given above.

LEMMA 5.4. *Let* F *and* S *be geodesic subspaces of* X *with* $S \subset T_v(F)$. *Then* S *is a parallel translate of a subspace of* F. *If, moreover,* F *and* S *are* r-*flats, then* $F = S$.

Proof. Let $\pi: X \to F$ denote the orthogonal projection onto F. Let L be a geodesic line in S. The function $p \to d(p,F)$ is convex on every line by Lemma 3.6 and is bounded on L. Therefore it is constant. For any points p and q on L, the quadrilateral $p\pi(p)\,\pi(q)q$ lies in a flat geodesic subspace E. Let g denote the element in $pol\,E$ sending $\pi(p)$ into p. Then g sends $\pi(q)$ into q. Since L and q are arbitrary in S, we find that $\pi(S)$ is a geodesic subspace, $g\pi(S) = S$, and $g \in Z(pol\,\pi(S))$. This proves the first assertions.

The second assertion follows at once from the observation that $Z(pol\,F)$ keeps F invariant if F is an r-flat since $N(pol\,F) = G_F$ by Lemma 5.1.

LEMMA 5.5. *Let* E *and* F *be geodesic subspaces with* pol E *and* pol F *commuting element-wise. Let* $x \in E \cap F$, *and let* D *denote the union of all geodesic lines in* E *perpendicular to* F *at* x. *Then* D *is a geodesic subspace.*

Proof. Let X and Y be in the Lie algebra of $pol\,E$ and let Z be in the Lie algebra of $pol\,F$. Then $0 = \mathrm{Tr}\,[XY]Z + \mathrm{Tr}\,X[ZY] = \mathrm{Tr}\,[XY]Z$;

that is, Z is orthogonal to the commutator subalgebra of *pol* E. Without loss of generality we may assume that group G is self-adjoint and x is fixed by $G \cap O(n, \mathbf{R})$. This implies that *pol* E and *pol* F are self-adjoint groups and that the subset of $\dot{pol}$ E which is orthogonal to $\dot{pol}$ F is a subalgebra $\mathcal{H}$. Let H denote the analytic group with Lie algebra $\mathcal{H}$. It is easy to see that H is a self-adjoint analytic group and $Hx_0 = D$. It follows by (3.4) that D is a geodesic subspace.

§6. A Basic Inequality

LEMMA 6.1. *Let* A *be an abelian analytic subgroup in* $P(n,R)$ *and let* $a(t)$ *be a differentiable path in* A *with* $a(0) = 1$. *Let* $y(t)$ *be a differentiable path in* $P(n,R)$. *Set*

$$\begin{aligned} 2Y(t) &= \log y(t) \\ H &= 2\dot{a}(0) \\ p(t) &= a(t)\, y(t)\, a(t)\ . \end{aligned}$$

Then at $t = 0$,

$$\mathrm{Tr}\,(p^{-1}\dot{p})^2 = \mathrm{Tr}\,((\cosh \operatorname{ad} Y)(H) + (\sinh \operatorname{ad} Y/\operatorname{ad} Y)(2\dot{Y}))^2\ .$$

Proof. Differentiating $p(t)$, we get

$$\dot{p} = \dot{a}\,y\,a + a\,\dot{y}\,a + a\,y\,\dot{a}\ .$$

At $t = 0$, we have $p(0) = y(0)$ and thus at $t = 0$,

$$\begin{aligned} p^{-\frac{1}{2}}\dot{p}\,p^{-\frac{1}{2}} &= y^{-\frac{1}{2}}\dot{a}\,y^{\frac{1}{2}} + y^{-\frac{1}{2}}\dot{y}\,y^{-\frac{1}{2}} + y^{\frac{1}{2}}\dot{a}\,y^{-\frac{1}{2}} \\ &= (e^{\operatorname{ad} Y} + e^{-\operatorname{ad} Y})(\dot{a}) + \tau_{2Y}(w\dot{Y}) \end{aligned}$$

where $\tau_Y(\dot{Y}) = (\exp - Y/2)(\exp\dot{} \; Y)(\exp - Y/2)$. By Lemma 3.1, $\tau_{2Y}(2\dot{Y}) = ((\sinh \operatorname{ad} Y)/\operatorname{ad} Y)(2\dot{Y})$. Since $\mathrm{Tr}\,(p^{-1}\dot{p})^2 = \mathrm{Tr}\left(p^{-\frac{1}{2}}\dot{p}\,p^{-\frac{1}{2}}\right)^2$, the lemma is now evident.

LEMMA 6.2. *Let* G *be a semi-simple analytic linear group, let* X *denote the associated symmetric Riemannian space, let* F *be a flat subspace of* X, *and let* $\pi : X \to F$ *denote the orthogonal projection of* X *onto* F. *Let* $p \in X$, *let* K *denote the stabilizer in* G *of the point* $\pi(p)$, *and let* $\mathcal{P}$ *denote the orthogonal complement in the Lie algebra of* G *to the Lie subalgebra of* K. *Let* Y *denote the unique element in* $\mathcal{P}$ *such that*

$$\exp Y\,(\pi(p)) = p\ .$$

Let $j : \mathcal{P} \to \dot{X}_{\pi(p)}$ *denote the canonical map of* $\mathcal{P}$ *onto the tangent space to* X *at* $\pi(p)$. *Set* $f(t) = (t \cosh t/\sinh t)^{\frac{1}{2}}$. *Then for any tangent vector* $C \in \dot{X}_p$,

$$|C| \geq |jf(\mathrm{ad}\ Y)\,j^{-1}\dot{\pi}_p(C)|\ .$$

Proof. We can assume that G is a self-adjoint subgroup of $GL(n, R)$ (cf. 2.6). Since G operates transitively on X and both sides of the inequality above are invariant under isometries, no generality is lost in assuming $K = G \cap O(n, R)$. Then $\mathcal{P} = S(n, R) \cap G$. We identify X with a subset of $P(n, R)$ via the map $\mu : gK \to g\,{}^t g$. Thereby $\pi(p)$ is identified with the identity matrix, F is identified with the abelian analytic subgroup $A = pol\, F$, and the projection $\pi : X \to F$ becomes identified with the map $aya \to a^2$ where $a \in A$ and $y \in A^{\perp}$, the union of geodesics through 1 orthogonal to A. By definition of Y, we have $\exp Y \cdot 1 \cdot \exp Y = p$. Given a tangent vector C to X at p, we select a differentiable path $p(t)$ with $p(0) = p$, $\dot{p}(0) = C$. There is a unique differentiable path $y(t)$ in $A^{\perp}$ such that

$$p(t) = a(t)\, y(t)\, a(t)\ .$$

Define $Y(t)$ by the relation $\exp 2\,Y(t) = y(t)$ with $Y(t) \in S(n, R)$, and set $\dot{Y} = \dot{Y}(0)$, $H = \dot{\pi}_p(C)$. Then $H = \dot{a}^2(0) = 2\,\dot{a}(0)$, and $|C|^2 = \mathrm{Tr}(p^{-1}\dot{p})^2$. Applying Lemma 6.1, we get

$$|C|^2 = \mathrm{Tr}\,((\cosh \mathrm{ad}\ Y)(H) + ((\sinh \mathrm{ad}\ Y)/\mathrm{ad}\ Y)(2\dot{Y}))^2\ .$$

The linear map $(\mathrm{ad}\, Y)^2$ stabilizes $\mathcal{P}$ and is self-adjoint with respect to $\mathrm{Tr}\, XY$ (cf. Section 3). Let $\eta_1, \ldots, \eta_m$ be an orthonormal set of eigenvectors for $(\mathrm{ad}\, Y)^2$ and let $v_1^2, \ldots, v_m^2$ denote the corresponding eigenvalues. Set $c_i = \cosh v_i$, $s_i = \sinh v_i / v_i, i = 1, \ldots, m$, and $H = \sum_1^m A_i \eta_i$, $2Y = \sum_1^m B_i \eta_i$. Then

$$|C|^2 = \mathrm{Tr}\left(\sum_1^m (c_i A_i + s_i B_i)\eta_i\right)^2$$
$$= \sum_1^m (c_i A_i + s_i B_i)^2 .$$

Rearrange the indices so that

$$A_i B_i \geq 0 \qquad i = 1, \ldots, h$$

$$A_i B_i < 0 \qquad i = h+1, \ldots, m .$$

Set $e_i = c_i A_i + s_i B_i$ and $w_i = c_i / s_i$ $(i = 1, \ldots, m)$. Then $B_i = s_i^{-1}(e_i - c_i A_i)$ and thus

$$|C|^2 \geq \sum_1^h c_i^2 A_i^2 + \sum_1^h s_i^2 B_i^2 + 2 \sum_1^h A_i B_i + \sum_{h+1}^m e_i^2$$
$$\geq \sum_1^h c_i^2 A_i^2 + \sum_1^h s_i^2 B_i^2 - 2 \sum_{h+1}^m A_i B_i + \sum_{h+1}^m e_i^2$$

since $\sum_1^m A_i B_i = 2\, \mathrm{Tr}\, H \dot{Y} = 0$. Writing

$$-2 \sum_{h+1}^m A_i B_i = -2 \sum_{h+1}^m s_i^{-1} A_i (e_i - c_i A_i)$$

we get

$$|C|^2 \geq \sum_1^h c_i^2 A_i^2 + s_i^2 B_i^2 + \sum_{h+1}^m \left(2 s_i^{-1} c_i A_i^2 - 2 s_i^{-1} e_i A_i + e_i^2\right)$$
$$\geq \sum_1^h c_i^2 A_i^2 + s_i^2 B_i^2 + \sum_{h+1}^m s_i^{-1} c_i A_i^2$$
$$+ \sum_{h+1}^m s_i^{-1} c_i \left(A_i^2 - 2 c_i^{-1} e_i A_i + c_i^{-1} s_i e_i^2\right)$$
$$\geq \sum_1^m s_i^{-1} c_i A_i^2 + \sum_{h+1}^m s_i^{-1} c_i \left(A_i^2 - 2 c_i^{-1} e_i A_i + c_i^{-2} e_i^2\right)$$

inasmuch as $s_i c_i^{-1} \geq c_i^{-1} \geq c_i^{-2}$ since $s_i \geq 1$ and $c_i \geq 1$. Consequently

$$|C|^2 \geq \sum_1^m w_i A_i^2 \sum_{h+1}^m w_i(A_i - c_i^{-1} e_i)^2 \geq \sum_1^m w_i A_i^2 .$$

Clearly $|jf(\text{ad } Y)j^{-1} \dot{\pi}_p(C)| = \left|\sum_1^m w_i^{\frac{1}{2}} A_i \eta_i\right|$. Therefore

$$|C| \geq |jf(\text{ad } Y)j^{-1} \dot{\pi}_p(C)| .$$

REMARK 1. In case the stabilizer of a point q in X is $G \cap O(n, R)$, we can compose the maps $\mathcal{P} \xrightarrow{j} \dot{X}_q \xrightarrow{\mu_q} \mathcal{P}$ and we see that $\dot{\mu}_q \circ j(Y) = 2Y$ since $\exp Y(q) = \exp Y \cdot \mu(q) \cdot \exp Y$.

REMARK 2. Inasmuch as $f(t) = (t \cosh t/\sinh t)^{\frac{1}{2}}$ is an even function of t, $f(t)$ is a power series in t^2. Since $(\text{ad } Y)^2$ keeps $\mathcal{P}$ stable, the map $jf(\text{ad } Y)j^{-1}$ is thus a well defined endomorphism of $\dot{X}_{\pi(p)}$. Indeed it is expressible in terms of the curvature tensor since $R(Y_1, Y_2, Y_3) = [[Y_1, Y_2], Y_3]$ for any elements Y_1, Y_2, Y_3 in $\mathcal{P}$. However, in order to compare $|C|$ and $|\dot{\pi}_p(C)|$, it is convenient to consider ad Y rather than $(\text{ad } Y)^2$ and to restate Lemma 6.2 in a slightly different way.

Let $M(n, R)$ and $S(n, R)$ denote respectively the set of all real $n \times n$ matrices and all symmetric real $n \times n$ matrices. On $M(n, R)$ we introduce the inner product $\langle U, V\rangle = \text{Tr } U\, {}^tV$. We recall that for any $Y \in S(n, R)$, ad Y is self-adjoint on $M(n, R)$ (cf. Section 3). An equivalent reformulation of Lemma 6.2 is

LEMMA 6.2 bis. *Let* π, p, Y, *and* C *be as in Lemma* 6.2. *Assume* G *is self-adjoint in* $GL(n, R)$ *and* $K = G \cap O(n, R)$. *Let* $\eta_1, \ldots, \eta_{n^2}$ *be an orthonormal base in* $M(n, R)$ *consisting of eigenvectors for* ad Y. *Then*

$$|C|^2 \geq \sum_{i=1}^{n^2} w_i A_i^2$$

where $[Y, \eta_i] = v_i \eta_i$, $w_i = |v_i \cosh v_i (\sinh v_i)^{-1}|$ $(i=1, \ldots, n^2)$, *and*

$$2\, j^{-1}(\dot{\pi}_p(C)) = \sum_1^{n^2} A_i \eta_i.$$

Proof. This follows immediately from Lemma 6.2.

For any tangent vector C to the symmetric Riemannian space X, we have denoted by $|C|$ the length of C under the invariant metric induced by the embedding μ of X into $P(n,R)$. For any real $n \times n$ matrix Y, we set

$$\|Y\| = (\mathrm{Tr}\, Y\, {}^tY)^{\frac{1}{2}} .$$

For any $Y \in \mathcal{P}$, we have $\mu_1(j(Y)) = 2Y$ and thus $\|2Y\| = |j(Y)| = d(\exp Y K, K) = d(p, \pi(p))$.

LEMMA 6.3. *Let* π, p, Y, *and* C *be as in Lemma* 6.2. *Assume that the semi-simple group* G *is self-adjoint in* $GL(n,R)$ *and* $K = G \cap O(n,R)$. *Set* $H = \dot{\pi}_p(C)$. *Then, identifying* H *with* $\dot{\mu}_1(H)$,

$$\frac{|C|}{|H|} \geq (2n)^{-\frac{1}{4}} \left\| \left[\frac{Y}{\|Y\|}, \frac{H}{\|H\|} \right] \right\| \cdot \|Y\|^{\frac{1}{2}} .$$

Proof. By Lemma 6.2 bis, $|C|^2 \geq \sum w_i A_i^2$, where $w_i = |v_i \cosh v_i (\sinh v_i)^{-1}|$ and $[Y, \eta_i] = v_i\, \eta_i$ $(i = 1, \ldots, n^2)$. Since $H = \sum A_i\, \eta_i$ and $\eta_1, \ldots, \eta_{n^2}$ is orthonormal, we have $\|H\|^2 = \sum A_i^2$.

As is well-known, $\mathrm{Tr}\, Z = 0$ for all $Z \in \dot{G}$. Let $\lambda_1, \ldots, \lambda_n$ denote the eigenvalues of the element Y. Then $\sum \lambda_i = \mathrm{Tr}\, Y = 0$ and

$$\mathrm{Tr}\,(\mathrm{ad}\, Y)^2 = \sum_{i,j=1}^{n} (\lambda_i - \lambda_j)^2 = 2n \sum_{i=1}^{n} \lambda_i^2 = 2n\, \mathrm{Tr}\, Y^2 .$$

Therefore, $\|Y\|^2 = (2n)^{-1}\, \mathrm{Tr}\,(\mathrm{ad}\, Y)^2 = (2n)^{-1} \sum_{i=1}^{n^2} v_i^2$. Set $u_i = |v_i|/((2n)^{\frac{1}{2}} \|Y\|)$ $(i = 1, \ldots, n^2)$. Inasmuch as $w_i \geq |v_i|$, we have

$$|C|^2 \geq \sum w_i A_i^2 \geq \sum |v_i|\, A_i^2 = (2n)^{\frac{1}{2}} \|Y\| \sum u_i A_i^2 \geq (2n)^{\frac{1}{2}} \|Y\| \sum u_i^2 A_i^2$$

since $u_i^2 \leq u_i \leq 1$. Hence $|C|^2 \geq (2n)^{-\frac{1}{2}} \|Y\|^{-1} \sum v_i^2 A_i^2$. But

$\|[Y,H]\|^2 = \| \sum_1^{n^2} v_i A_i \eta_i \|^2 = \sum v_i^2 A_i^2$. Thus

$$|C|^2 \geq (2n)^{-\frac{1}{2}} \|Y\|^{-1} \|[Y,H]\|^2 \ .$$

By definition, $|H| = \|\dot{\mu}_1(H)\| = \|H\|$. Consequently

$$\frac{|C|}{|H|} \geq (2n)^{-\frac{1}{4}} \left\| \left[\frac{Y}{\|Y\|} , \frac{H}{\|H\|} \right] \right\| \|Y\|^{\frac{1}{2}} \ .$$

It is of course desirable to free Lemma 6.3 of the hypothesis that $K = G \cap O(n,\mathbf{R})$. Accordingly, we introduce the following notation.

For any $g \in GL(n,\mathbf{R})$, and any $Y \in M(n,\mathbf{R})$, set

$$\theta_g(Y) = g^t(g^{-1} Y g)g^{-1} = g\, {}^t g\, {}^t Y\, {}^t g^{-1}\, g^{-1} \ ,$$

$$\|Y\|_g^2 = \operatorname{Tr} Y\, \theta_g(Y) \ .$$

Suppose now that G is a self-adjoint analytic subgroup of $GL(n,\mathbf{R})$. Let $\mathfrak{o}$ denote the point in the associated symmetric space X that is stabilized by $G \cap O(n,\mathbf{R})$ and let $q = h\mathfrak{o}$ with $h \in G$. Set

$$J_q(g) = gq \ , \qquad \text{if } g \in G$$

$$j_q(Y) = \dot{J}_q(Y) \ , \qquad \text{if } Y \in \dot{G} \cap h^{-1}S(n,\mathbf{R})h \ .$$

With these notations, we can equally well restate Lemma 6.2 bis and Lemma 6.3 with less restricted hypotheses.

LEMMA 6.3 bis. *Let* G *be a self-adjoint semi-simple analytic group, let* X *denote the associated symmetric space, let* F *be a flat subspace of* X, *and let* $\pi : X \to F$ *denote the orthogonal projection of* X *onto* F. *Let* $p \in X$, *let* K *denote the stabilizer of* $\pi(p)$ *in* G, *let* $\mathcal{P}$ *denote the orthogonal complement in* $\dot{G}$ *to* $\dot{K}$, *and let* Y *denote the unique element in* $\mathcal{P}$ *such that*

$$\exp Y(\pi(p)) = p \ .$$

Choose g *in* G *with* $g\mathit{o} = \pi(p)$. *Let* $C \in \dot{X}_p$, *set* $H = \dot{\pi}_p(C)$, *and* $H' = 2\, j^{-1}_{\pi(p)}(H)$. *Let* $\eta_1, \ldots, \eta_{n^2}$ *be an orthonormal set of eigenvectors for* $\operatorname{ad} Y$ *on* $M(n, \mathbf{R})$ *with respect to the inner product* $\operatorname{Tr} U\, \theta_g(V)$. *Set*

$$H' = \sum_1^{n^2} A_i \eta_i,\ [Y, \eta_i] = v_i \eta_i,\ w_i = |v_i \cosh v_i (\sinh v_i)^{-1}|\ (i = 1, \ldots, n^2)\ .$$

Then

$$|C|^2 \geq \sum_1^{n^2} w_i\, A_i$$

and $2^{-1} d(p, \pi(p)) = \|Y\|_g = (2n)^{-1} \quad v_i^2$. *Moreover*

$$\frac{|C|}{|H|} \geq (2n)^{-\frac{1}{4}} \left\| \left[\frac{Y}{\|Y\|_g}, \frac{H'}{\|H'\|_g} \right] \right\|_g \|Y\|_g^{\frac{1}{2}} \ .$$

LEMMA 6.4. *Let* X *be a simply-connected symmetric Riemannian space of rank* r *and of negative curvature (i.e., its isometry group has no compact or vector normal subgroups). Let* F *be an* r*-flat in* X *and let* $\pi : X \to F$ *denote the orthogonal projection of* X *onto* F. *Let* $p \in X$ *and let* T *be an* r*-dimensional subspace of the tangent space to* X *at* p. *Let* τ *denote the restriction of* $\dot{\pi}_p$ *to* T. *Then*

$$\det \tau \leq c\, d(p, F)^{-\frac{1}{2}}$$

where c *is a constant depending only on the space* X.

Proof. No generality is lost in assuming G is self-adjoint in $GL(n, \mathbf{R})$ and that the stabilizer of $\pi(p)$ is $G \cap O(n, \mathbf{R})$.

The map τ takes the unit ball in T into an ellipsoid and $\det \tau$ is, up to a constant depending only on r, the product of the principal axes of the ellipsoid. Since π is a projection, the longest axis of the ellipsoid has length at most 1. On the other hand, the shortest principal axis must be no longer than the length of a radius of the ellipsoid along some *regular* geodesic issuing from the center of the ellipsoid.

We may assume that $\det \tau \neq 0$, otherwise there is nothing to prove. In the case $\det \tau \neq 0$, set $C = \tau^{-1}(H)$. We have $|C| = 1$, and $\|2Y\| = d(p, \pi(p)) = d(p, F)$ where $\exp Y \cdot \pi(p) = p$ and Y is chosen as in Lemma 6.2. By Lemma 6.3

$$|H| \leq 2^{\frac{1}{2}} (2n)^{\frac{1}{4}} \left\| \left[\frac{Y}{\|Y\|}, \frac{H}{\|H\|} \right] \right\|^{-1} \|2Y\|^{-\frac{1}{2}} .$$

Therefore $|H| \leq (8n)^{\frac{1}{4}}\, c(H)\, d(p, F)^{-\frac{1}{2}}$, where

$$c(H)^{-1} = \inf \{ \|[Y/\|Y\|, H/\|H\|]\| \,;\, Y \in (\mathit{pol}\ F)^{\perp} \cap \mathcal{P} \} .$$

Inasmuch as H lies along a regular geodesic of F, the centralizer of H in $\mathcal{P}$ lies in the Lie algebra of *pol* F. Consequently, $c(H) < \infty$. From this, Lemma 6.4 follows at once.

§7. Geometry of Neighboring Flats

In this section we shall determine the intersection of an r-flat F with a tubular neighborhood $T_V(F_0)$ of an r-flat F_0. The principal result (Theorem 7.8) states that the intersection is approximately an intersection of half-spaces with singular faces.

We continue the notation and assumptions of Section 4.

Let F be a flat subspace of X. In Section 5 we have defined *pol* F as the unique maximal polar subgroup of the stabilizer G_F. Suppose now S denotes either a geodesic ray, or a chamber wall or a chamber in X, and let F denote the unique minimal flat subspace of X containing S. Clearly the stabilizer of S lies in G_F, and we may define *pol* S as the stabilizer of S in *pol* F. Let ◄S denote the intersection of all chambers and chamber walls containing S. Then *pol* S ⊂ *pol* ◄S. Now *pol* ◄S is a chamber or chamber wall in G (cf. Section 4 and 2.4) and P(*pol* ◄S) is a well-defined parabolic subgroup of G. We set

$$P(S) = P(pol\ ◄S)\ .$$

LEMMA 7.1. *Let* S *denote a geodesic ray or a chamber wall or a chamber in* X. *Then*

(i) $hd(S, gS) < \infty$ *if and only if* $g \in P(S)$;

(ii) $d(S, gS) = 0$ *if* $g \in U(S)$.

Proof. If S is a chamber in X, then the assertion is a restatement of Lemma 4.1. Indeed, the proof of Lemma 4.1 applies equally well in case S is a ray or a chamber wall. Namely, we have $S = (pol\ S)x$ where x is the origin of S, and for any $p \in P(S)$, $\{a^{-1}pa;\ a \in pol\ S\}$ is a bounded set. Hence

$$\begin{aligned} hd(pS, S) &\leq \sup\{d(pax, ax);\ a \in pol\ S\} \\ &\leq \sup\{d(a^{-1}pax, x);\ a \in pol\ S\} \\ &< \infty . \end{aligned}$$

Given now an element g in G, we write $g = pk$ with $p \in P(S)$ and k in the isotropy subgroup of x. Then

$$\begin{aligned} hd(gS, S) &\geq hd(pkS, pS) - hd(pS, S) \\ &\geq hd(kS, S) - hd(pS, S) \\ &= \infty , \end{aligned}$$

unless $kS = S$, (cf. proof of Lemma 4.1). But $kS = S$ implies $k \in G_S \subset Z(pol\ S) = Z(pol\ {}^{\blacktriangleleft}S) \subset P(S)$. Lemma 7.1 (i) now follows. Assertion (ii) follows from the fact that $\lim_{a \to \infty} a^{-1} g a = 1$ for $g \in U(S)$.

LEMMA 7.2. (i) *Let* L_0 *and* L *be rays in* X. *Then* $hd(L_0, L) < \infty$ *if and only if* $L_0 = gL$ *with* $g \in P(L)$.

(ii) *Let* ${}^{\blacktriangleleft}S_0$ *and* ${}^{\blacktriangleleft}S$ *be chamber walls in* X. *Then* $hd({}^{\blacktriangleleft}S_0, {}^{\blacktriangleleft}S) < \infty$ *if and only if* ${}^{\blacktriangleleft}S_0 = g\,{}^{\blacktriangleleft}S$ *with* $g \in P({}^{\blacktriangleleft}S)$.

Proof. Let ${}^{\blacktriangleleft}F_0^*$ be a closed chamber containing L_0 whose origin is the origin of L_0 and let ${}^{\blacktriangleleft}F^*$ be a closed chamber similarly related to L. Then there is a $g \in G$ such that $g\,{}^{\blacktriangleleft}F^* = {}^{\blacktriangleleft}F_0^*$, since G operates transitively on the set of chambers of X. Therefore there is a ray L_1 in ${}^{\blacktriangleleft}F$ such that $gL_1 = L_0$. By (2.7) $G = KP(L_1)$ where K is any maximal compact subgroup of G; we take K to be the isotropy subgroup of the origin of ${}^{\blacktriangleleft}F$. Set $g = kp$ with $k \in K$ and $p \in P(L_1)$. Then $hd(L_1, pL_1) < \infty$ by Lemma 7.1.

Suppose now that $hd(L_0, L) < \infty$. Then

$$\begin{aligned} hd(kL_1, L) &\leq hd(kL_1, kpL_1) + hd(kpL_1, L) \\ &\leq hd(L_1, pL_1) + hd(L_0, L) < \infty . \end{aligned}$$

It follows at once that $kL_1 = L$ since both are rays with the same origin. However, both L_1 and L belong to the same closed chamber and by (vii) of (2.3) if $kak^{-1} = b$ with a and b in the same closed chamber of a polar subgroup, then $a = b$ and $k \in Z(a)$. It follows at once that $L_1 = L$ and $k \in Z(\mathit{pol}\ L)$. Thus $L_0 = gL_1 = kpL$ with $p \in P(L)$ and $k \in Z(\mathit{pol}\ L) \subset P(L)$. Hence $g \in P(L)$. This proves the "only if" assertion (i) of Lemma 7.2. The other assertion follows from the previous lemma.

To prove (ii), choose a geodesic ray L_0 in the interior of $\blacktriangleleft S_0$ having the same origin as $\blacktriangleleft S_0$. Set $c = hd(\blacktriangleleft S_0, \blacktriangleleft S)$. Then $\blacktriangleleft S_0 \subset T_{2c}(\blacktriangleleft S)$ implies $L_0 \subset T_{2c}(T_{2c}(L_0) \cap \blacktriangleleft S)$. Hence $T_{2c}(L_0) \cap \blacktriangleleft S$, which is convex by Lemma 3.5 contains a ray L having the same origin as $\blacktriangleleft S$. It is easy to see that $hd(L, L_0) < \infty$. Hence by (i), $L_0 = gL$; with $g \in P(L)$. Since L_0 is in the interior of $\blacktriangleleft S$ it follows that L is in the interior of S. By definition therefore, $P(L) = P(\blacktriangleleft S)$. Proof of (ii) is now complete.

DEFINITION. Let L be a geodesic ray in X. A geodesic ray of the form gL with $g \in Z(\mathit{pol}\ L)$ is called a *parallel translate* of L.

Since $Z(\mathit{pol}\ L) = Z(\mathit{pol}\ \tilde{L})$ where $\tilde{L}$ is the geodesic line containing L, we see that the definition of parallel translate is compatible with the definition of parallel translate of flat subspaces given in Section 5.

The next lemma is a basic step towards understanding how portions of two flats can remain within finite distance of each other as we go off towards infinity.

LEMMA 7.3. *Let* F *be a flat subspace of* X *and let* L_0 *be a geodesic ray in* $T_v(F)$. *Then*

(i) L_0 *approaches a parallel translate of a geodesic ray* L *in* $F \cap T_v(L_0)$; *that is, there is a ray* L *in* $F \cap T_v(L_0)$ *and a* $z \in Z(\mathit{pol}\ L)$ *such that* $d(L_0, zL) = 0$.

(ii) *Any ray in* $T_v(L_0) \cap F$ *is parallel to* L.

(iii) *If* L_0 *is a regular geodesic ray and* F *is an* r-*flat, then* $d(L_0, F) = 0$.
(iv) *If* F_0 *and* F *are* r-*flats and* $F_0 \subset T_v(F)$, *then* $F_0 = F$.

Proof. $L_0 \subset T_v(F)$ implies that $L_0 \subset T_v(T_v(L_0) \cap F))$. Therefore $T_v(L_0) \cap F$ is an unbounded set which is convex by Lemma 3.5. Let L be a geodesic ray in $T_v(L_0) \cap F$. Clearly $hd(L_0, L) < \infty$. By Lemma 7.2, $L_0 = gL$ with $g \in P(L)$. We write $g = uz$ with $z \in Z(pol\, L)$, $u \in U(L)$. Then zL is parallel to L. Moreover $U(zL) = zU(L)z^{-1} = U(L)$. Thus $d(zL, uzL) = 0$ by Lemma 7.1, since $u \in U(zL)$. This proves (i).

Assertion (ii) follows from the fact that two rays in F at finite Hausdorff distance are parallel.

To prove (iii), we assume that L_0 is regular. Then L is regular since $L = g^{-1} L_0$. It follows at once that $Z(pol\, L) = Z(pol\, F)$ where F is the unique r-flat containing L. Since $Z(pol\, F)$ stabilizes F, we find $zL \subset F$ and thus $d(L_0, F) \leq d(L_0, zL) = 0$.

Proof of (iv): By Lemma 3.7,

$$d(p, F) = d(F_0, F)$$

for all $p \in F_0$. By (iii), $d(F_0, F) = 0$. Hence $d(p, F) = 0$ for all $p \in F_0$. Therefore $F_0 \subset F$. Since F_0 is an r-flat, $F_0 = F$.

LEMMA 7.4. *Let* $\blacktriangleleft F$ *and* $\blacktriangleleft F_0$ *be chambers in* X *and let* $\blacktriangleleft S$ *be a wall of* $\blacktriangleleft F$. *Assume* $\blacktriangleleft S \subset T_v(\blacktriangleleft F_0)$. *Then* $\blacktriangleleft F_0 = g \blacktriangleleft F$ *with* $g \in P(\blacktriangleleft S)$. *Moreover, if* L *is a geodesic ray in* $\blacktriangleleft F \cap T_v(\blacktriangleleft F_0)$ *and* L_0 *is a ray in* $\blacktriangleleft F_0 \cap T_v(\blacktriangleleft L)$ *then there is a* $g_0 \in P(L)$ *such that* $g_0 L = L_0$ *and* $g_0 \blacktriangleleft F = \blacktriangleleft F_0$.

Proof. Let L be a geodesic ray in the interior of the chamber wall $\blacktriangleleft S$ and having the same origin. Then $T_v(L) \cap \blacktriangleleft F_0$ is convex by Lemma 3.5 and is unbounded. Choose a ray L_0 in $T_v(L) \cap \blacktriangleleft F_0$. Then $hd(L_0, L) < v$. Let L'_0 be the ray parallel to L_0 with origin that of $\blacktriangleleft F_0$.

Then $hd(L, L'_0) < \infty$. Hence by Lemma 7.2, $L'_0 = pL$ with $p \in P(L)$. By choice of L, $\blacktriangleleft S$ is the intersection of all chamber walls containing L. Hence $P(L) = P(\blacktriangleleft S)$. Moreover, $p\blacktriangleleft S$ is the wall of the chamber $\blacktriangleleft F_0$ containing the ray L'_0. Since all the chambers with wall $p\blacktriangleleft S$ are conjugate under $Z(pol\ p\blacktriangleleft S) = p\,Z(pol\ \blacktriangleleft S)p^{-1} \subset P(\blacktriangleleft S)$ (cf. 2.3 (vii)), we conclude that $\blacktriangleleft F_0 = g\blacktriangleleft F$ with $g \in P(\blacktriangleleft S)$. This proves the first assertion.

The proof of the second assertion is essentially contained in the proof above, the only alteration being that all the closed chambers containing the ray pL are conjugate under $Z(pol\ pL) = p\,Z(pol\ L)p^{-1} \subset P(L)$.

COROLLARY 7.4.1. *Let* $\blacktriangleleft S$ *and* $\blacktriangleleft S_0$ *be chamber walls in* X *and let* $\blacktriangleleft S \subset T_v(\blacktriangleleft S_0)$. *Then there is a* $g \in P(\blacktriangleleft S)$ *with* $g\blacktriangleleft S$ *a wall of* $\blacktriangleleft S_0$.

Proof. Let $\blacktriangleleft F$ and $\blacktriangleleft F_0$ be chambers in X whose closures contain $\blacktriangleleft S$ and $\blacktriangleleft S_0$ respectively. Applying Lemma 7.4, we find a $g \in P(\blacktriangleleft S)$ with $g\blacktriangleleft S$ a wall of $\blacktriangleleft F_0$. By Lemma 7.1, $hd(\blacktriangleleft S, g\blacktriangleleft S) < \infty$. Hence $g\blacktriangleleft S \subset T_w(\blacktriangleleft S_0)$ for some finite w. Since $\blacktriangleleft S_0$ and $g\blacktriangleleft S$ are walls of $\blacktriangleleft F_0$, we infer that $g\blacktriangleleft S$ is a wall of $\blacktriangleleft S_0$.

Let F and F_0 be r-flats, and let $x \in F$. Let $\mathcal{R}$ denote the set of all rays in F having origin x and contained in $T_v(F_0)$ for some v.

DEFINITION. $$F_0 \cap_x F = \bigcup \{L;\ L \in \mathcal{R}\}\ .$$

LEMMA 7.5. *Let* $g \in G$, *and let* F *be an* r-*flat in* X, *and let* $x \in F$. *Let* $\mathcal{S}$ *denote the set of all chambers and chamber walls in* F *of origin* x. *Then*

(i) $gF \cap_x F$ *is the convex hull of* $\{\blacktriangleleft S;\ \blacktriangleleft S \in \mathcal{S},\ g \in P(\blacktriangleleft S)G_F\}$

(ii) $gF \cap_x F \subset T_v(gF) \cap F$ *if* $v > d(gF, x)$.

Proof. Set $F_0 = gF$ and let L be a ray in F of origin x and contained in, $T_v(F_0)$. Since $T_v(L) \cap F_0$ is an unbounded convex set, it contains a ray L_0. Clearly, $hd(L_0, L) < \infty$. By Lemma 7.4, $L_0 = g_0 L$ and $F_0 = g_0 F$ and $g_0 \in P(L)$. Consequently, $F = g_0^{-1} gF$ and $g \in g_0 G_F \subset P(L)G_F$. That is, if $L \subset gF \cap_x F$, then $g \in P(L)G_F$. Conversely, if $g \in P(L)$, then $hd(gL, L) < \infty$ by Lemma 7.1 and as a result $L \subset gF \cap_x F$ for any $g \in P(L)G_F$. In other words $gF \cap_x F = \bigcup \{L;\ g \in P(L)G_F, L \text{ of origin } x\}$.

On the other hand, one sees directly from the definition of $gF \cap_x F$ that it is convex; for it is the union over v of the set of all rays of origin x in the convex set $T_v(gF) \cap F$.

For any ray L, there is a unique chamber or chamber wall $\blacktriangleleft S$ containing L in its interior and having the same origin, and one has $P(\blacktriangleleft S) = P(L)$. Therefore

$$gF \cap_x F = \bigcup \{\blacktriangleleft S;\ \blacktriangleleft S \in \mathcal{S},\ g \in P(\blacktriangleleft S)G_F\}$$

or equally well $gF \cap_x F$ is the convex hull of $\{\blacktriangleleft S;\ \blacktriangleleft S \in \mathcal{S},\ g \in P(\blacktriangleleft S)G_F\}$. Thus (i) is proved.

By definition, for any ray L in $gF \cap_x F$ of origin x, $L \subset T_v(gF)$ for some v. Consider the distance from a variable point in L to gF. This function is bounded on L and is convex by Lemma 3.6. Since a bounded convex function on a half-line is monotonically decreasing, it takes on its maximum at its endpoint. Hence $d(y, gF) \le d(x, gF)$ for all $y \in L$. From this assertion (ii) follows.

LEMMA 7.6. *Let* $x \in P(n, R)$, *let* u *be a unipotent triangular matrix in* $GL(n, R)$, *and let* D *denote the diagonal subgroup of* $GL(n, R)$. *Then*

(i) $$\log^2 n^{-1} \operatorname{Tr} x \le \operatorname{Tr} \log^2 x \le n \log^2 \operatorname{Tr} x$$

(ii) $$d(u \cdot 1, D) \ge -2 \log n + (4n)^{-\frac{1}{2}} d(u \cdot 1, 1)$$

where $d(\ ,\ )$ *denotes the invariant metric on* $P(n, R)$ *and* $u \cdot 1 = u\,{}^t u$.

Proof. Let $\xi_1, \ldots, \xi_n$ denote the eigenvalues of x, and set $\xi_{max} = \max_i \xi_i$. Then

$$\log^2 \mathrm{Tr}\, x = \log^2 \Sigma_i\, \xi_i \leq \log^2 n\, \xi_{max} \leq \mathrm{Tr} \log^2 n\, x\ .$$

Replacing x by $n^{-1}x$ yields the first half of (i). From

$$\mathrm{Tr} \log^2 x = \Sigma_i \log^2 \xi_i \leq n \log^2 \xi_{max} \leq n \log^2 \mathrm{Tr}\, x$$

the second half follows.

If we replace x by x^{-1} in the first half of (i) and note that $\log^2 x = \log^2 x^{-1}$, we get $\mathrm{Tr} \log^2 x \geq \log^2 n^{-1} \mathrm{Tr}\, x^{-1}$ and therefore

$$(\mathrm{Tr} \log^2 x)^{\frac{1}{2}} \geq \frac{1}{2} (\log n^{-1} \mathrm{Tr}\, x + \log n^{-1} \mathrm{Tr}\, x^{-1})$$

$$\geq -\log r + \frac{1}{2} (\log \mathrm{Tr}\, x + \log \mathrm{Tr}\, x^{-1})\ .$$

Set $(c_{ij}) = u$ and let $\lambda \in D$. Then

$$d(u \cdot 1, \lambda^{-1}) = d(\lambda u \cdot 1, 1)$$

$$= (\mathrm{Tr} \log^2 \lambda u\, {}^t(u\lambda))^{\frac{1}{2}} \geq -\log n + \frac{1}{2} (\log \mathrm{Tr}\, x + \log \mathrm{Tr}\, x^{-1})$$

where $x = (\lambda u)\, {}^t(\lambda u)$. Clearly $\mathrm{Tr}\, x = \sum_{i,j} \lambda_i^2\, c_{ij}^2$ and $\mathrm{Tr}\, x^{-1} \geq \lambda_i^{-2}$ for each i where $\lambda = \mathrm{diag}\,(\lambda_1, \ldots, \lambda_n)$. Consequently,

$$d(u \cdot 1, \lambda^{-1}) \geq -\log n + \frac{1}{2} \left(\log \lambda_i^2\, c_{ij}^2 - \log \lambda_i^2\right)$$

$$\geq -\log n + \frac{1}{2} \log c_{ij}^2\ .$$

Set $c_{max} = \sup_{i,j} c_{ij}^2$. Then $n^2\, c_{max} \geq \mathrm{Tr}\, u\, {}^tu$ and therefore $\log c_{max} \geq \log n^{-2}\, \mathrm{Tr}\, u\, {}^tu = -2 \log n + \log \mathrm{Tr}\, u\, {}^tu$. By the right half of (i), $\log \mathrm{Tr}\, u\, {}^tu \geq n^{-\frac{1}{2}}.(\mathrm{Tr} \log^2 u\, {}^tu)^{\frac{1}{2}}$. Therefore

$$d(u \cdot 1, \lambda^{-1}) \geq -\log n + \frac{1}{2} \left(-2 \log n + n^{-\frac{1}{2}}\, d(u \cdot 1, 1)\right)$$

$$\geq -2 \log n + (4n)^{-\frac{1}{2}}\, d(u \cdot 1, 1)\ .$$

LEMMA 7.7. *Let* $\blacktriangleleft S$ *be a chamber or chamber wall of origin* p *in the* r*-flat* F. *Suppose* g *is an element in* G *not in* $P(\blacktriangleleft S)G_F$. *Then for any positive* v, (i) $\blacktriangleleft S \cap T_v(gF)$ *lies within a bounded distance of a wall of* $\blacktriangleleft S$; (ii) $\blacktriangleleft S \cap T_v(gF) \subset T_t(gF \cap_p F)$ *for some finite* t.

Proof. Let $a \in pol \blacktriangleleft S$. By the Bruhat decomposition, (2.8), we can write $g^{-1} = vwu$ with v in the unipotent radical of $P(\blacktriangleleft S)$, $w \in N(pol\, F) = G_F$, and $u \in P(\blacktriangleleft S)$. Hence

$$d(gF, ap) = d(F, g^{-1} ap) = d(F, vwuap)$$
$$= d(F, w^{-1} vwuap)$$

since $w^{-1}F = F$. We have $w^{-1}vw \in v_- P(\blacktriangleleft S)$ with v_- in the unipotent radical U of $P(-\blacktriangleleft S)$, where $-\blacktriangleleft S$ denotes the opposite of $\blacktriangleleft S$. Moreover $v_- \neq 1$ since $g^{-1} \in G_F P(\blacktriangleleft S)$. Thus $w^{-1}vwu = v_- u'$ with $u' \in P(\blacktriangleleft S)$ and $d(gF, ap) = d(F, v_- u'ap) = d(F, a^{-1} v_- aq)$, where $q = a^{-1} u'ap$. As a varies over $pol \blacktriangleleft S$, the element q remains bounded; set $r = \sup\{d(p,q);\, a \in pol \blacktriangleleft S\}$. Since orthogonal projection onto F diminishes distance

$$-d(F, a^{-1} v_- aq) + d(F, a^{-1} v_- ap) \leq 2(d(a^{-1} v_- aq,\, a^{-1} v_- ap)) \leq 2r\ .$$

We may select a faithful representation of G so that the stabilizer of the point p is $G \cap O(n, \mathbf{R})$, and that moreover the groups U and $pol\, F$ are respectively triangular and diagonal. Then without loss of generality, we may assume that $d(a^{-1} v_- ap, p)^2 = \mathrm{Tr}\, \log^2 (a^{-1} v_- a)\, {}^t(a^{-1} v_- a)$. Let $(c_{ij}) = v_-$ and $\mathrm{diag}(\lambda_1, \ldots, \lambda_n) = a$. Then

$$d(a^{-1} v_- ap, p) = (\mathrm{Tr}\, \log^2 (a^{-1} v_- a)\, {}^t(a^{-1} v_- a))^{\frac{1}{2}}$$
$$\geq \log n^{-1}\, \mathrm{Tr}\, (a^{-1} v_- a)\, {}^t(a^{-1} v_- a)$$
$$\geq -\log n + \log \sum_{i,j} \left(\lambda_i\, c_{ij}\, \lambda_j^{-1}\right)^2 .$$

Inasmuch as $v_- \neq 1$, there is a fundamental root α which is positive on the Lie algebra of *pol* $\blacktriangleleft S$ such that v_- has a non-zero component in the root space of $-\alpha$; that is, $\alpha(a) = \log \lambda_i \lambda_j^{-1}$ with $c_{ij} \neq 0$. Thus

$$d(a^{-1} v_- a p, p) \geq (-\log n + \log c_{ij}^2) + 2\alpha(a) .$$

By Lemma 7.6, $d(a^{-1} v_- a p, F) \geq -2 \log n + (4n)^{-\frac{1}{2}} d(a^{-1} v_- a p, p)$. Putting together the inequalities above, we get

$$\begin{aligned} d(gF, a p) &= d(F, a^{-1} v_- a q) \geq d(F, a^{-1} v_- a p) - 2r \\ &\geq c\, d(a^{-1} v_- a p, p) - c' \\ &\geq 2c\alpha(a) - c'' \end{aligned}$$

where $c = (4n)^{-\frac{1}{2}}$, $c' = 2r + 2 \log n$, $c'' = c' + c(\log n - \log c_{ij}^2)$. Thus $ap \in T_v(gF)$ implies $2x\alpha(a) - c'' < v$; that is, $\alpha(a) < (2c)^{-1}(v + c'')$. This implies assertion (i) of Lemma 7.7. Let $R = \blacktriangleleft S \cap (gF \cap_p F)$. Clearly R is the largest wall of $\blacktriangleleft S$ which lies in $gF \cap_p F$. By applying assertion (i) of Lemma 7.7 repeatedly, we find a sequence of chamber walls $S_0, S_1, \ldots, S_n$ with $S_0 = \blacktriangleleft S, S_i$ a wall of S_{i-1}, $S_n = R$, and for each $v > 0$, $S_{i-1} \cap T_v(gF) \subset T_{v_i}(S_i)$ for some v_i $(i = 1, \ldots, n)$. From this chain of inclusions, we find for any positive v

$$\blacktriangleleft S \cap T_v(gF) \subset T_t(R)$$

for some finite t. This proves assertion (ii).

THEOREM 7.8. *Let* F *and* F_0 *be* r*-flats in* X, *let* $x \in F$ *and let* $s > d(x, F_0)$. *Then*

$$F_0 \cap_x F \subset T_s(F_0) \cap F \subset T_t(F_0 \cap_x F)$$

where t *is a number depending on* F_0, F, x *and* s.

Proof. By (ii) of Lemma 7.5, $F_0 \cap_x F \subset T_s(F_0) \cap F$ for $s > d(x, F_0)$. Let $\blacktriangleleft S$ be a chamber or chamber wall in F of origin x. Let g be an element of G such that $gF = F_0$. By Lemma 7.5 (i), $\blacktriangleleft S \subset F_0 \cap_x F$ if and only if $g \in P(\blacktriangleleft S)G_F$. By Lemma 7.7, $\blacktriangleleft S \cap T_s(F_0) \subset T_t(F_0 \cap_x F)$ for some finite t if $\blacktriangleleft S \not\subset F_0 \cap_x F$. Inasmuch as F is the union of all the $\blacktriangleleft S$, and there are only a finite number of chambers and chamber walls in F of origin x, we see that there is a finite t such that $F \cap T_s(F_0) \subset T_t(F_0 \cap_x F)$.

LEMMA 7.9. *Let X be a symmetric space of constant negative curvature, and let F and F_0 be geodesic lines in X with $d(F, F_0) > 0$. Then $T_v(F) \cap F_0$ has length at most $2v/(1-(\cosh d_0)^{-1})$, where $d_0 = d(F, F_0)$ and d is a suitably normalized invariant metric.*

Proof. We adopt the notation of Lemma 6.2. Let π denote the orthogonal projection of X on F. Then for any tangent vector C to F_0 at p, $|C|^2 = \mathrm{Tr}\,(\cosh \mathrm{ad}\, Y(H) + (\sinh \mathrm{ad}\, Y/\mathrm{ad}\, Y)(2\dot{Y}))^2$ where H is the projection of C onto $\dot{F}_{\pi(p)}$, and Y is orthogonal to $\dot{F}_{\pi(p)}$ with $\exp Y(\pi(p)) = p$. In a space of constant curvature, we have $(\mathrm{ad}\, Y)^2\, \dot{F}_{\pi(p)} \subset \dot{F}_{\pi(p)}$ and $(\mathrm{ad}\, Y)^2\, \dot{F}^T_{\pi(p)} \subset \dot{F}^T_{\pi(p)}$ for any $Y \in \dot{F}^T_{\pi(p)}$. Therefore

$$\begin{aligned} |C|^2 &= \mathrm{Tr}\,((\cosh \mathrm{ad}\, Y(H)) + ((\sinh \mathrm{ad}\, Y/\mathrm{ad}\, Y)(2\dot{Y})^2)) \\ &\geq \mathrm{Tr}\,(\cosh \mathrm{ad}\, Y(H))^2 = T_y\,(\cosh^2 \alpha(Y))\,(H)^2 \end{aligned}$$

where $(\mathrm{ad}\, Y)^2(H) = \alpha(Y)^2 H$. Now in the metric of X induced from the adjoint representation, of the group G of isometries of X $d(p, \pi(p))^2 = \mathrm{Tr}\,(\mathrm{ad}\, Y)^2 = (\dim G/G_F)\alpha(Y)^2$. Thus multiplying the distance by $(\dim G/G_F)^{-1}$, we get an invariant metric with $|\alpha(Y)| = d(p \cdot \pi(p)) = d(p, F)$ and therefore $|C|/|H| \geq \cosh d(p, F) \geq \cosh d_0$. Let s denote the length of $T_v(F) \cap F_0$ and let t denote the length of $\pi(T_v(F) \cap F_0)$. Then $s \leq 2v + t \leq 2v + s/\cosh d_0$. Hence

$$s \leq 2v/(1-(\cosh d_0)^{-1}) .$$

§8. Density Properties of Discrete Subgroups

LEMMA 8.1 (Selberg). *Let* G *be a locally compact group and let* Γ *be a discrete subgroup.*

(i) *If* G/Γ *is compact, then for all* $\gamma \in \Gamma$, $Z(\gamma)/Z(\gamma) \cap \Gamma$ *is compact, where* $Z(\gamma)$ *denotes the centralizer of* γ.

(ii) *If* G/Γ *has finite measure, then for any* $g \in G$ *and for any neighborhood* U *of* 1 *in* G, *there is a positive integer* n *such that* $Ug^nU \cap \Gamma \neq \emptyset$.

Since the proof is very short, we present it.

(i) Let $\gamma \in \Gamma$. Consider the map $k: G \to G$ given by $k(g) = g\gamma g^{-1}$. Then k is continuous. Since $k(\Gamma) \subset \Gamma$, $k(\Gamma)$ is closed and accordingly, $k^{-1}(k(\Gamma))$ is closed; that is, $\Gamma Z(\gamma)$ is closed in G. Therefore $\Gamma\backslash\Gamma Z(\gamma)$ is a closed subset of the compact space $\Gamma\backslash G$ and is compact. Hence $Z(\gamma)/Z(\gamma) \cap \Gamma$ is compact.

(ii) Since left translation is a measure-preserving map of G/Γ, we can find two distinct positive integers k and ℓ such that $g^kU\Gamma \cap g^{\ell}U\Gamma \neq \emptyset$. Hence $u^{-1}g^{k-\ell}U \cap \Gamma \neq \emptyset$. Applying this remark to $U \cap U^{-1}$, assertion (ii) follows.

LEMMA 8.2. *Let* A *be a maximal polar subgroup of the semi-simple group* G. *Let* V *and* W *denote neighborhoods of the identity in* A *and* G *respectively. Let* $c > 1$ *and set*

$$A_c = \{a \in A;\ \alpha(a) > c \text{ for all positive } \mathbf{R}\text{-roots } \alpha\}.$$

Then there is a neighborhood U *of the identity in* G *such that for any* $a \in A_c$

$$U a U \subset W[M a V]$$

where $g[x]$ *denotes* $g x g^{-1}$ *and* M *is the maximum compact subgroup of* $Z(A)$.

This result is proved in my paper "On intersections of Cartan subgroups with discrete subgroups," Indian Journal of Mathematics, Vol. 34 (1970), 203-214.

Let Γ be a discrete subgroup of the semi-simple group G. Let X denote the symmetric space associated to G and let $r = \text{rank } X$. Let F be an r-flat in X.

DEFINITION. F is Γ-compact if and only if $\Gamma \backslash \Gamma F$ is compact.

LEMMA 8.3. *Let* Γ *and* X *be as above. Assume that* G/Γ *is compact. Then the set of* Γ*-compact* r*-flats is dense in the set of all* r*-flats of* X.

Proof. Let F be an r-flat in X, and let A be the maximum polar subgroup of G_F, i.e., $A = \mathit{pol}\, F$. Let g be a polar regular element in A. Then $g \in A_c$ for some $c > 1$ and some ordering of the R-roots on A.

Let V be a neighborhood of 1 in A such that $A_c V \subset A_d$ with $d > 1$. Let W be a neighborhood of 1 in G. Select a neighborhood U of 1 in G such that $U A_c U \subset W[M A_c V]$. Finally, by Selberg's Lemma, we can find an element $\gamma \in \Gamma \cap U A_c U$. By Lemma 8.2, $\gamma \in w[M A_d]$ with $w \in W$ and $d > 1$. Hence γ is polar regular. Clearly, $\gamma \in w X(A) w^{-1} \in w G_F w^{-1} = G_{wF}$. Furthermore, $Z(\gamma)$ operates transitively on wF by Lemma 5.2 (iii). Inasmuch as $Z(\gamma)/Z(\gamma) \cap \Gamma$ is compact and $Z(\gamma) = G_{wF}$, it follows at once that wF is Γ-compact.

We have proved: Given any neighborhood W of the identity in G and any r-flat F in X there is a Γ-compact r-flat of the form wF with $w \in W$. Thus, Lemma 8.3 is proved.

REMARK. Lemma 8.3 is valid for discrete subgroups Γ of a semi-simple group G such that G/Γ has finite measure. To see this, one needs an adequate condition for $Z(\gamma)/Z(\gamma) \cap \Gamma$ to be compact. Such a condition has been given by Raghunathan.

Raghunathan defines an element $g \in G$ to be R-*hyper-regular* if and only if the number of eigenvalues counted with multiplicity of modulus 1 of $\Lambda(g)$ is as small as possible and -1 is not an eigenvalue of $\Lambda(g)$, where $\Lambda(g)$ is the representation of G on the exterior algebra over $\dot{G}$, the Lie algebra of G.

Raghunathan's criterion. If γ is R-hyper-regular, then $Z(\gamma)/Z(\gamma) \cap \Gamma$ is compact.

Let $b \in A_c$ be an R-hyper-regular element and let V be a neighborhood of 1 in A such that $A_c V \subset A_d$ with $d > 1$ and $\bigcup_{j=1}^{\infty} b^j V$ consists entirely of R-hyper-regular elements. Select a neighborhood U of 1 in G such that for all $a \in A_c$, $U a U \subset W[M a V]$. By Selberg's Lemma, we can find an element $\gamma \in \Gamma$ $U b^j U$ for some $j > 0$. Since $U b^j U \subset W[M b^j V]$, we find $\gamma \in w[M b^j V]$ with $w \in W$. Hence γ is R-hyper-regular. Since an R-hyper-regular element is polar regular (cf. [13], Remark 1.2; note that their "R-regular" is our "polar regular") and since $Z(\gamma)/Z(\gamma) \cap \Gamma$ is compact, our proof applies to the case that G/Γ has finite measure. Thus, modulo the proof of Raghunathan's criterion (which is proved in [13]), we have proved

LEMMA 8.3′. *Let* Γ *and* X *be as above. Assume that* G/Γ *has finite measure. Then the set of* Γ*-compact* r*-flats is dense in the set of all* r*-flats of* X.

LEMMA 8.4 (Mautner). *Let* G *be a semi-simple analytic group having no compact normal subgroup of positive dimension, and let* A *be a semi-group of polar regular elements in* G *(which is contained in no proper normal subgroup of* G*). Let* Γ *be a closed subgroup of* G *such that* G/Γ *has finite* G*-invariant measure. Then*

(i) A *operates ergodically on* G/Γ.

(ii) $\Gamma x A$ *is dense in* G *for almost all* $x \in G$.

Assertion (i) was proved by Mautner for polar subgroups A in [11]. Assertion (ii) follows immediately from (i). Details may be found in [12g].

LEMMA 8.5. *Let* G *be a semi-simple analytic group having no compact normal subgroup of positive dimension, and let* P *be a parabolic subgroup. Let* Γ *be a closed subgroup such that* G/Γ *has finite* G*-invariant measure. Then* $\overline{\Gamma P} = G$.

Proof. No generality is lost in assuming that P is a minimal parabolic subgroup. Let $\blacktriangleleft A$ be a chamber such that $P = P(\blacktriangleleft A)$ (cf. (2.4)). Hence $\Gamma u \blacktriangleleft A^{-1}$ is dense in G for almost all u in G. Given $g \in G$, we can find sequences of elements $\{\gamma_n\}$ in Γ and $\{a_n\}$ in $\blacktriangleleft A$ such that $\gamma_n u_n^{-1} a_n^{-1} \to g$ and $u_n \to 1$ as $n \to \infty$. Thus $\gamma_n = w_n g a_n u_n$ with $w_n \to 1$ as $n \to \infty$.

Set $N = U(\blacktriangleleft A^{-1})$, the unipotent subgroup generated by the root spaces of the negative roots on $\blacktriangleleft A$. Then $u_n = v_n p_n$ with $v_n \in N$ and $p_n \in P$. Thus

$$\gamma_n P = w_n g a_n v_n P = w_n g(a_n v_n a_n^{-1})P .$$

We have $a_n v_n a_n^{-1} \to 1$ as $n \to \infty$. Hence $\gamma_n P \to gP$ as $n \to \infty$. Therefore $\overline{\Gamma P} = G$.

LEMMA 8.6 (Borel). *Let* G *be a semi-simple linear analytic group having no compact normal subgroup of positive dimension, and let* Γ *be a closed subgroup such that* G/Γ *has finite* G*-invariant measure. Then* Γ *is Zariski-dense in* G.

This lemma is proved by Borel in [2a] and, generalizes a result first proved by Selberg (cf. [16]) in the case $G = SL(n, \mathbf{R})$. Another proof is given in [12f].

§9. Pseudo-Isometries

Let X and X' be metric spaces and let $\phi : X \to X'$ be a continuous map. Let k and b be positive numbers.

DEFINITION. ϕ is a (k,b) *pseudo-isometry* if and only if

(9.1.1) $$d(\phi(x), \phi(y)) \le k\, d(x,y), \quad \text{for all } x, y \text{ in } X$$

and

(9.1.2) $$d(\phi(x), \phi(y)) \ge k^{-1} d(x,y), \quad \text{if } d(x,y) \ge b\ .$$

The map ϕ is called a *pseudo-isometry* if it is a (k,b) pseudo-isometry for some (k,b). If, for example, X is compact and ϕ satisfies a Lipschitz condition with constant k, then ϕ is a (k,b) pseudo-isometry where b is the diameter of X. Thus the condition that ϕ be a pseudo-isometry is not much of a restriction unless X is non-compact.

Let $\phi : X \to X'$ be a (k,b) pseudo-isometry, and let $S \subset X$. For any tubular neighborhood of radius r, we get from (9.1.1)

(9.1.1)′ $$\phi(T_r(S)) \subset T_{kr}\phi(S)\ .$$

From (9.1.2) we get

(9.1.2)′ $$\phi^{-1}(T_r(\phi(S)) \subset T_{r'}(S),\ r' = \sup(kr, b)\ .$$

A continuous map $\phi : X \to X'$ satisfying only (9.1.2) is called (k,b)-*incompressible*.

We shall require the following well-known fact from the theory of fiber bundles.

LEMMA 9.1. *Let* X *and* X' *be contractible topological spaces on which the group* Γ *operates freely; that is* $gx \neq x$ *for any* $x \in X$ *and* $g \in \Gamma$ *unless* $g = 1$. *Assume that* $\Gamma\backslash X$ *and* $\Gamma\backslash X'$ *are finite simplicial complexes. Then there is a* Γ*-space morphism* $\phi: X \to X'$ *such that the induced map* $\bar{\phi}: \Gamma\backslash X \to \Gamma\backslash X'$ *is simplicial. Moreover* $\bar{\phi}$ *is a homotopy equivalence.*

Proof may be found in [17].

The next lemma is central for our method.

LEMMA 9.2. *Let* G *and* G' *be semi-simple analytic groups, let* K *and* K' *be maximal compact subgroups in* G *and* G' *respectively, and set* $X = G/K$, $X' = G'/K'$. *Let* Γ *and* Γ' *be torsion-free discrete subgroups such that* G/Γ *and* G'/Γ' *are compact. Let* $\theta: \Gamma \to \Gamma'$ *be an isomorphism. Then there is a pseudo-isometry* $\phi: X \to X'$ *such that*

$$\phi(\gamma x) = \theta(\gamma)\phi(x)$$

for all $\gamma \in \Gamma$ *and* $x \in X$; *that is,* ϕ *is a* Γ*-space morphism and a pseudo-isometry.*

Proof. Observe first that Γ operates freely on X. For if G_x denotes the stabilizer of a point x in X, then G_x is conjugate to K and therefore $\Gamma \cap G_x$ is discrete and compact. Since $\Gamma \cap G_x$ is thus finite, we have $\Gamma \cap G_x = (1)$ since Γ is torsion-free. Similarly, Γ' operates freely on X'.

The space $\Gamma\backslash X$ is a differentiable manifold and is compact since it is homeomorphic to $\Gamma\backslash G/K$. Therefore, it can be triangulated and is finite simplicial complex. Similarly $\Gamma'\backslash X'$ is a finite simplicial complex. One can now apply Lemma 9.1 and choose a Γ-space morphism $\phi: X \to X'$. It remains to show that ϕ is a pseudo-isometry.

Let $\bar{\phi}: \Gamma\backslash X \to \Gamma'\backslash X'$ denote the induced map of $\Gamma\backslash X$ to $\Gamma'\backslash X'$. Then $\bar{\phi}$ is a simplicial map by our choice. We regard $\Gamma\backslash X$ as a metric space with metric induced from X. Similarly for $\Gamma'\backslash X'$. Being a simplicial map of a finite simplicial complex, $\bar{\phi}$ satisfies a Lipschitz condition with some constant $k_1 : d(\bar{\phi}(x), \bar{\phi}(y)) \le k_1\, d(x,y)$ for all x, y in $\Gamma\backslash X$. Inasmuch as X and X' are covering spaces of $\Gamma\backslash X$ and $\Gamma'\backslash X'$ and have the covering metrics, it follows at once that

$$d(\phi(x), \phi(y)) \le k_1\, d(x,y)$$

for all x and y in $\Gamma\backslash X$.

It remains to prove the opposite inequality. Let $\pi : X \to \Gamma\backslash X$ and $\pi' : X' \to \Gamma'\backslash X'$ denote the natural projections.

Let S denote the set of all closed geodesics in the metric space $\Gamma'\backslash X'$ whose length does not exceed some positive number s and which are not homotopic to a point. Then S is a compact space when topologized by the Hausdorff metric for subsets of $\Gamma'\backslash X'$. If S is non-empty, the geodesic length function must attain its minimum on S; that is, S contains a closed geodesic of minimum length s_0. Set $r' = s_0/2$. We assert: *for all* $\gamma' \in \Gamma'$ *and* $x' \in X$ *with* $\gamma' \neq 1$,

$$d(x', \gamma' x') \ge 2r' \,. \tag{9.2.1}$$

Proof. By (5.3.2) and the remarks following it, $\inf\{d(x', \gamma' x');\ x' \in X'\} = \inf\{d(x', \gamma' x');\ x' \in F'\}$ where F' is a geodesic line stable under γ'. Here we are using that every element of Γ' is semi-simple. This implies that $\inf\{d(x', \gamma' x');\ x' \in X'\}$ is given by the length of a closed geodesic whose image in the fundamental group of $\Gamma'\backslash X'$ is γ' when we identify the fundamental group with Γ'. Consequently, $\inf\{d(x', \gamma' x');\ \gamma' \in \Gamma', \gamma' \neq 1, x' \in X'\} \ge s_0$. From this (9.2.1) follows.

As a direct consequence of (9.2.1), π' is injective on any ball in X' of radius less than r'.

Next we assert: *Let* B' *be a closed ball in* X' *of diameter* r'. *Then*

$$\operatorname{diam} \phi^{-1}(B') < \infty \,. \tag{9.2.2}$$

Proof. From the commutative diagram

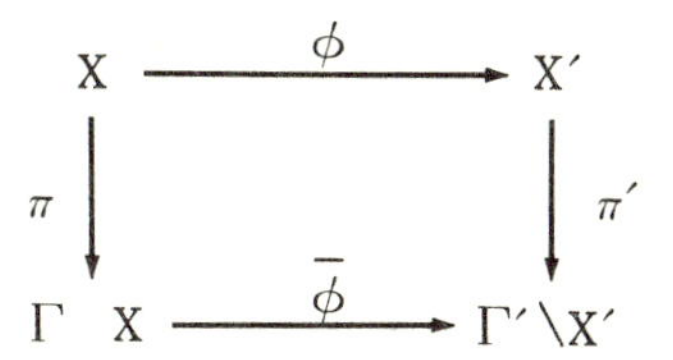

we see that $\pi\phi^{-1}(B') = \bar{\phi}^{-1}\pi'(B')$. Since $\bar{\phi}$ is a proper map and $\pi'(B')$ is compact it follows that $\pi\phi^{-1}(B')$ is closed in $\Gamma\backslash X$ and in fact compact. Consequently $\Gamma\phi^{-1}(B') = \pi^{-1}(\pi\phi^{-1}(B'))$ is closed in X. Set $B = \phi^{-1}(B')$.

The map ϕ stretches distance by at most k_1. Hence there is a positive number δ such that for any x, y in $\Gamma\backslash X$, $d(x,y) < \delta$ implies $d(\bar{\phi}(x), \bar{\phi}(y)) < r'/4$. It follows easily that $\phi(T_\delta(B)) \subset T_{r'/4}(\phi(B)) \subset T_{r'/4}(B')$, where $T_{r'/4}(B')$ is a ball of radius less than r'. We know that π' is injective on $T_{r'/4}(B')$. Inasmuch as ϕ is a Γ-space morphism, it follows at once for any element γ in Γ that

$$\gamma T_\delta(B) \cap T_\delta(B) \neq \emptyset \text{ implies } \gamma = 1\ . \tag{9.2.3}$$

Suppose for the moment that $\operatorname{diam} \phi^{-1}(B') = \infty$. Then there is an unbounded sequence of elements in B. We have seen above that $\pi(B)$ is compact. Therefore $\Gamma\backslash\Gamma B$ is compact and we can select a subsequence $\{b_1, b_2, \ldots, b_n, \ldots\}$ such that $b_n \to \infty$ and $\Gamma b_n \to \Gamma b$ with $b \in B$; that is $\gamma_n b_n \to b$ with $\gamma_n \in \Gamma$ and $b_n \to \infty$.

Choose n so that $d(\gamma_n b_n, b) < \delta$. Then $\gamma_n b_n \in T_\delta(B)$ and $b_n \in B$; that is $\gamma_n T_\delta(B) \cap T_\delta(B) \neq \emptyset$. It follows from (9.2.3) that $\gamma_n = 1$. Therefore $d(b_n, b) < \delta$ for all large n – a contradiction. This establishes (9.2.2).

Modulo Γ', the closed balls of diameter r' in X' make up a compact space $\mathcal{B}$. Set

$$r = \sup_{B'} \operatorname{diam} \phi^{-1}(B')$$

as B' ranges over all balls in X' of diameter r'. Then $r < \infty$ since $\operatorname{diam} \phi^{-1}(B')$ is an upper semi-continuous function on $\mathcal{B}$. Set $k_2 = r/r'$. Then given any x and y in X with $d(\phi(x), \phi(y)) < n r'$, there is a path in X joining x to y of length less than $n r$ for any positive integer n. Given now x and y in X, we choose n so that $n r' \leq d(\phi(x), \phi(y)) < (n+1) r'$. Then

$$d(x, y) < (n+1) r = \frac{n+1}{n} \frac{r}{r'} n r' \leq \frac{n+1}{n} k_2 \, d(\phi(x), \phi(y)) .$$

Set $k = \sup(k_1, 2k_2)$ and $b = r$. Then ϕ is a (k, b) pseudo-isometry.

§10. Pseudo Isometries of Simply Connected Spaces with Negative Curvature

LEMMA 10.1. *Let* $\phi: R^m \to R^n$ *be a* (k,b) *incompressible map. Then* $n \geq m$. *If* $n \leq m$, *then* $n = m$ *and* ϕ *is surjective.*

Proof. Let B_r (resp. B'_r) denote the ball in R^m (resp. R^n) with center at the origin 0 and radius r. Without loss of generality we can assume that $\phi(0) = 0$.

On the boundary ∂B_r of B_r select $m+1$ equally spaced points and join each pair by a great circle arc. Thereby we get a curvilinear triangulation of the closure of B_r as an m-simplex.

Let $S_1, \ldots, S_{m+1}$ denote the m-1 dimensional faces of $\partial \overline{B}_r$. Let $S_{i,s} = \{tx;\ s \leq t \leq 1,\ x \in S_i\}$. Then $\{\overline{B}_{sr}, S_{i,s}, i = 1, \ldots, m+1\}$ is a closed cover C_s of the pair $\{\overline{B}_r, \overline{B}_{sr}\}$. Let N denote the nerve of the cover C_s and let M denote the subcomplex of N whose supports meet ∂B_r. Then the canonical map of the cohomology groups $H^*(N,M)$ of the nerve to the Čech cohomology groups $H^*(\overline{B}_r, \partial B_r)$ of the pair $(\overline{B}_r, \partial B_r)$ is an isomorphism. (One appeals to standard theorems for simplicial complexes; alternatively, one can use Leray's spectral sequence for closed covers with acyclic supports, its applicability to the cohomology of a pair being deduced from the cohomology sequence of a pair.)

It should be remarked that the nerves M and N are independent of s for $0 < s \leq 1$; that M is isomorphic to the boundary of an m-simplex and N is isomorphic to the cone over M.

We next wish to approximate the map ϕ of $(\overline{B}_r, \partial B_r)$ into $(\phi(\overline{B}_r), \phi(\overline{B}_r) - B'_{r/k})$ via suitable open coverings.

Observe first that no generality is lost in assuming $b > 0$. Select a positive number e such that

$$ke = b \ .$$

Next, select $r > 2b$ such that

$$\phi(B_{2b}) \subset B'_{r/k} \ .$$

Since $r > b$, we have $\phi(\partial B_r) \subset R^n - B'_{r/k}$.

Set $s = 2b/r$; i.e., $rs = 2ke$. Let C' denote the open cover of $\partial(\overline{B}_\gamma)$ given by

$$C' = \{T_e(\phi(S));\ S \in C_s\} \ .$$

Let C denote the open cover of $\overline{B}_r$ given by

$$C = \{\phi^{-1}(S');\ S' \in C'\} \ .$$

Since ϕ is (k, b) incompressible and $ke \geq b$, we have for all $S \in C_s$,

$$\phi^{-1}(T_e(\phi(S)) \subset T_{ke}(S) \ .$$

Let $\hat{C} = \{T_{ke}(S);\ S \in C_s\}$.

Now by elementary geometry we see that:

(10.2). *the map* $S \to T_{ke}(S)$ *of the cover* C_s *to the cover* $\hat{C}$ *induces an isomorphism of nerves.*

Denoting by $N(\ )$ the nerve of the cover $(\ )$, we find that the composition

$$N(C_s) \to N(C) \to N(\hat{C})$$

is an isomorphism, where the arrow denotes the map induced by inclusion. It follows at once that $N(C)$ as well as C_s determines the cohomology of the pair $(\overline{B}_r, \partial B_r)$. Set $N = N(C)$ and $N' = N(C')$.

Let M (resp. M') denote the subcomplex of N (resp. N') whose supports meet ∂B_r (resp. $\partial(\overline{B}_r) - B'_{r/k}$). Consider the commutative diagram

$$\begin{array}{ccc} H^*(N,M) & \xleftarrow{\ \alpha\ } & H^*(N',M') \\ \beta\downarrow & & \downarrow \\ H^*(\bar{B}_r, \partial B_r) & \xleftarrow{\ \phi^*\ } & H^*(\phi(\bar{B}_r), \phi(\bar{B}_r) - B'_{r/k}) \end{array}$$

where the vertical arrows are the canonical maps into Čech cohomology and the horizontal maps are induced by the map ϕ. Inasmuch as α and β are isomorphisms, the map ϕ^* is surjective. In particular the map

$$\phi^m : H^m(\phi(\bar{B}_r), \phi(\bar{B}_r) - B'_{r/k}) \to H^m(\bar{B}_r, \partial B_r)$$

is non-zero and therefore $H^m(\phi(\bar{B}_r), \phi(\bar{B}_r) - B'_{r/k}) \neq 0$. Now $H^*(\phi(\bar{B}_r), \phi(\bar{B}_r) - B'_{r/k}) = H^*(\phi(\bar{B}_r) \cup cB'_{r/k}, cB_{r/k})$ where $cB'_{r/k}$ denotes $R^n - B'_{r/k}$; and from $H^m(\phi(\bar{B}_r) \cup cB'_{r/k}, cB'_{r/k}) \neq 0$ we infer that $m \leq n$ from the cohomological definition of dimension. By hypothesis, $n \leq m$; hence $n = m$. We can conclude at once that $\phi(\bar{B}_r) \supset B'_{r/k}$ – otherwise $\phi(\bar{B}_r)$ could be deformed into $R^n - B'_{r/k}$ and thus the restriction image $H^r(R^n, cB'_{r/k}) \to H^n(\phi(\bar{B}_r) \cup cB'_{r/k}, cB'_{r/k})$ is 0. From the exactness of the cohomology sequence of $((R^n, \phi(B_r) \cup cB'_{r/k}, cB'_{r/k})$, we would get the contradiction $H^{n+1}(R^n, \phi(\bar{B}_r) \cup cB'_{r/k}) \neq 0$. Consequently, $\phi(\bar{B}_r) \supset B'_{r/k}$. This holding for all large r, we see that ϕ is surjective. The proof of Lemma 10.1 is now complete.

Lemma 10.1 is valid more generally for a wider class of metric spaces.

Let X be a Riemannian space that is complete and simply connected. For any point $p \in X$ and for any tangent vector $u \in \dot{X}_p$, the tangent space to X at p, we let $\exp tu$ denote the geodesic curve with origin at p and whose tangent at p is u. Set $d_p(\exp u, \exp v) = |u - v|$ for all u and v in $\dot{X}_p$, where $|u|$ is the length defined on $\dot{X}_p$ by the Riemannian metric on X.

The following definition is equivalent with the usual definition of non-positive sectional curvature on a simply connected X.

DEFINITION. X has *non-positive curvature* if and only if for all $p \in X$ and for all u and v in $\dot{X}_p$,

$$d(\exp u, \exp v) \geq |u - v| \tag{10.2}$$

where d denotes the Riemannian distance.

Inasmuch as any two points in a complete Riemannian manifold are joined by a geodesic, the map $u \to \exp u$ of $\dot{X}_p$ to X is surjective for all $p \in X$ and by virtue of (10.2) is also injective. Condition (10.2) is clearly equivalent to:

For all p, q, r in X,

$$d_p(q, r) \leq d(q, r) \; . \tag{10.3}$$

By Lemma 3.2, the space X of Section 3 is an example of a space of non-positive curvature.

LEMMA 10.1′. *Let* X *and* X' *be simply connected complete Riemannian manifolds of non-positive sectional curvature with* $\dim X = m$ *and* $\dim X' = n$. *Let* $\phi : X \to X'$ *be a continuous map. Let* $\mathfrak{o} \in X$, *set* $B_r = T_r(\mathfrak{o})$ *and* $B'_r = T_r(\phi(\mathfrak{o}))$. *Assume that* ϕ *is* (k, b) *incompressible on* $B_r, r > 2b$, *and* $\phi(B_{2b}) \subset B'_{r/k}$. *Then* $n \geq m$. *If moreover* $n \leq m$, *then* $n = m$ *and* $\phi(\overline{B}_r) \supset B_{r/k}$.

The proof of this Lemma is verbatim like the proof of Lemma 10.1, with two points requiring clarification.

The first point is that the ball B_r is the same for both the Riemannian metric and the "Euclidean" metric $d_{\mathfrak{o}}$ based at $\mathfrak{o}$, and distances along rays from $\mathfrak{o}$ coincide for both metrics. Accordingly, we define the closed cover $\{S_{1,s}, \ldots, S_{m,s}; \overline{B}_{sr}\}$ with respect to the metric $d_{\mathfrak{o}}$. However, because $d_{\mathfrak{o}}(x, y) \leq d(x, y)$ for any points x, y in X, we find $T_b(S) \subset T_b^{\mathfrak{o}}(S)$ where $T_b^{\mathfrak{o}}(S)$ denotes the tube of radius b about S with respect to the Euclidean metric $d_{\mathfrak{o}}$. Therefore, the assertion (10.2) is valid for the tubular neighborhoods in the given Riemannian metric. These two remarks having been made, the rest of the proof establishes Lemma 10.1′.

§11. Polar Regular Elements in Co-Compact Γ

Let G be a semi-simple analytic linear group and let Γ be a *discrete subgroup such that* G/Γ *is compact.* It is well-known that each element of Γ must be *semi-simple.*

(*Proof.* Let $\gamma \in \Gamma$. Let $\gamma = s \cdot u$ be the Jordan decomposition of γ with s semi-simple and u unipotent (cf. 2.1). Then Z(s), the centralizer of s in G, is reductive by (2.6) and therefore by (2.9) u lies in a subgroup of Z(s) locally isomorphic to $SL(2,\mathbf{R})$ if $u \neq 1$. From this we see that the closure of the conjugacy class $G[\gamma]$ contains the closure of $Z(s)[su]$ and thus contains s. But G/Γ compact and $\Gamma[\gamma]$ closed implies that $G[\gamma]$ is closed. Hence γ is conjugate to s and is semi-simple.)

The following Lemma is also known.

LEMMA 11.1. *Let* G *be a semi-simple analytic linear group and let* Γ *be a discrete subgroup such that* G/Γ *is compact. Let* $r = \mathbf{R}$*-rank* G. *Then*

(i) *Any abelian subgroup of* Γ *has rank at most* r.

(ii) *Any abelian subgroup* Δ *of* Γ *of rank* r *contains a polar regular element. In fact, the elements of* Δ *are polar regular except for those lying in a finite union* Ω *os subgroups of rank less than* r.

Proof. Let Δ be an abelian subgroup of Γ. Since each element of Δ is semi-simple, the group Δ is simultaneously diagonalizable over the complex numbers and therefore Δ is a reductive abelian group. Let Δ^* denote the intersection of G with all the algebraic groups containing Δ.

Then Δ^* is an abelian Lie group with a finite number of connected components (cf. (2.5)). Hence $\Delta^* = M \cdot A$ where M is the maximum compact subgroup of Δ^* and A is a polar subgroup (cf. (2.1)). Now $\Delta M/M = \Delta/\Delta \cap M$ is isomorphic to a discrete subgroup of the polar group A. Hence rank Δ = rank $\Delta/\Delta \cap M \leq \dim A \leq \mathbf{R}$-rank G, proving (i).

Suppose now that rank $\Delta = r$. Then $\dim A = r$ and A is a maximal polar subgroup of G. Since $(\Delta M) \cap A$ is a lattice in A, $\Delta M \cap A$ is not contained in the union W of the chamber walls of A. Hence $\Delta M \cap (A - W) \neq \emptyset$.

Set $\Omega = MW \cap \Delta$. Then $M\Omega \cap A \subset MW$. From $\Omega/\Omega \cap M = M\Omega/M \to W$ we deduce that Ω is a finite union of subgroups of rank at most $r-1$.

Suppose now that $\gamma \in \Delta - \Omega$. Then $\gamma = ma$ with $m \in M$ and $a \in A - M\Omega \subset A - MW \subset A - W$. Thus a is $\mathbf{R}$-regular. Since $ma = am$, we have $a = pol\,\gamma$ and thus γ is polar regular. Finally $\delta - \Omega$ is non-empty since $M(\Delta - \Omega) = M(M\Delta \cap (A-W)) \neq \emptyset$. This proves (ii).

The first assertion of the next lemma was first stated by J. A. Wolf in [22]. However, his proof is heuristic at a critical point and requires our Lemma 8.2 in order to be put in order.

LEMMA 11.2. *Let* G *be a semi-simple analytic linear group and let* Γ *be a discrete subgroup such that* G/Γ *is compact. Let* $r = \mathbf{R}$-rank G. *Then*

(i) Γ *contains a polar regular element.*

(ii) *If an element* γ *in* Γ *is polar regular, then* $Z(\gamma) \cap \Gamma$ *contains an abelian subgroup of finite index and rank* r.

Proof. The proof of (i) occurs in the proof of Lemma 8.3. Namely, we proved there: Given any maximal polar subgroup A in G and any chamber $\blacktriangleleft A$ in A, there is an element in Γ conjugate to an element in $M \blacktriangleleft A$, where M is the maximum compact subgroup in $Z(A)$, the centralizer of A in G. Since every element in $M \blacktriangleleft A$ is polar regular, assertion (i) follows.

Let γ be a polar regular element. Without loss of generality, we may assume that $\gamma \in M\ A$. Then

$$A \subset Z(\gamma) \subset Z(pol\ \gamma) = Z(A) = MA\ .$$

Hence $Z(A) = MZ(\gamma)$. Now $Z(\gamma)/Z(\gamma) \cap \Gamma$ is compact by Selberg's Lemma (8.1). Hence $Z(A)/Z(A) \cap \Gamma$ is compact. Inasmuch as M contains all compact subgroups of MA, the torsion subgroup of $MA \cap \Gamma$ is $M \cap \Gamma$. Since $MA \cap \Gamma/M \cap \Gamma = M(MA \cap \Gamma)/M \hookrightarrow A$, $MA \cap \Gamma/M \cap \Gamma$ is isomorphic to a lattice of the vector subgroup A and is therefore a free abelian group of rank r. In particular, there is a direct product splitting $MA \cap \Gamma = (M \cap \Gamma) \cdot \Delta$ with Δ a free abelian group of rank r. This proves Assertion (ii).

LEMMA 11.3. *Let Γ and Γ' be discrete subgroups of the semi-simple analytic groups G and G′ respectively such that G/Γ and G'/Γ' are compact. Let $\theta : \Gamma \to \Gamma'$ be an isomorphism and let Δ be an abelian subgroup of Γ of maximal rank. Then*

(i) *rank* Δ = $\mathbb{R}$*-rank* G.

(ii) $\theta(\Delta)$ *contains a polar regular element.*

(iii) $\mathbb{R}$*-rank* G = $\mathbb{R}$*-rank* G′.

Proof. Assertion (i) follows immediately from Lemma 11.1 (i) and Lemma 11.2 (ii).

Assertion (ii) follows immediately from Lemma 11.1 (ii) and Lemma 11.2.

Assertion (iii) follows immediately from Assertion (i).

REMARK. In the case that G/Γ has finite measure and is not necessarily compact, Lemma 11.3 (i) remains valid after the hypothesis that the elements of Δ are semi-simple, thanks to the criterion of Raghunathan for the compactness of $Z(\gamma)/Z(\gamma) \cap \Gamma$ (cf. Remark following Lemma 8.3). Instead of applying Selberg's criterion to polar regular elements as we did in proving Lemma 11.2 (ii), one applies Raghunathan's criterion to $\mathbb{R}$-hyperregular elements and reasons as above. The assertions of Lemma 11.1 become valid only after adding the hypothesis that the elements of the abelian subgroup are semi-simple. To sum up:

Let G *be a semi-simple linear analytic group and let* Γ *be a discrete subgroup such that* G/Γ *has finite measure. Let* $r = \mathbb{R}$*-rank* G. *Then*

(11.1.i)′. *Any abelian subgroup of semi-simple elements of* Γ *has rank at most* r.

(11.1.ii)′. *Any abelian subgroup* Δ *of* Γ *of rank* r *contains an* $\mathbb{R}$*-hyperregular element. In fact the elements of* Δ *are* $\mathbb{R}$*-hyperregular except for those lying in a finite union of subgroups of rank less than* r.

(11.2.i)′. Γ *contains an* $\mathbb{R}$*-hyperregular element.*

(11.2.ii)′. *If* $\gamma \in \Gamma$ *is* $\mathbb{R}$*-hyperregular, then* $Z(\gamma) \cap \Gamma$ *contains an abelian subgroup of semi-simple elements of finite index and rank* r.

(11.3.i)′. *Let* Δ *be an abelian subgroup of semi-simple elements of* Γ *of maximal rank. Then rank* $\Delta = \mathbb{R}$*-rank* G.

Finally, let (G', Γ') be a pair satisfying the same hypotheses as (G, Γ), and let $\theta : \Gamma \to \Gamma'$ is an isomorphism. Then

(11.3.ii)′. *If* $\theta(\Delta)$ *consists of semi-simple elements, then* $\theta(\Delta)$ *contains an* $\mathbb{R}$*-hyperregular element.*

(11.3.iii)′. *If* θ *carries semi-simple elements to semi-simple elements and vice-versa, then* $\mathbb{R}$*-rank* $G = \mathbb{R}$*-rank* G'.

In actual fact, the conclusion $\mathbb{R}$-rank $G = \mathbb{R}$-rank G' holds without the added hypothesis in (11.3.iii) about θ, but by a more involved argument than the one we used above. (cf. a forthcoming article by G. Prasad and M. S. Raghunathan [13].)

§12. Pseudo-Isometric Invariance of Semi-Simple and Unipotent Elements

We continue the notation G, K, X *of Section* 5.

In Lemma 5.3, we have seen that for any $g \in G$,

$$\inf_{x \in X} d(x, gx)^2 = 4 \operatorname{Tr} (\log pol\, g)^2 ;$$

moreover g is semi-simple or not according as the infimum of $d(x, gx)$ is *attained* on X or not. This property is a characterization of semi-simplicity that is invariant under isometries of X.

To get a criterion invariant under *pseudo-isometries*, we modify the above characterization.

For any real number a and for any element $g \in G$, we set

$$|g|^2 = 4 \operatorname{Tr} (\log pol\, g)^2$$

$$I(g, a) = \{x \in X;\ d(x, gx) \leq a\}$$

$$I(g) = I(g, |g|) .$$

As remarked above, g is a semi-simple element if and only if $I(g)$ is non-empty.

LEMMA 12.1.

(i) $I(g; a)$ *is a convex set.*

(ii) $I(g)$ *is a geodesic subspace.*

(iii) *If* g *is semi-simple, then* $Z(g)$ *operates transitively on* $I(g)$.

(iv) $x \in I(g)$ *if and only if* $Z(g)x$ *is a geodesic subspace.*

Proof. (i) Let x and y be points in $I(g; a)$. Let $x(s)$ denote the point on the geodesic line joining x to y with $d(x, x(s)) = s$. Set $f(s) = d(x(s), gx(s))$. Set $x_1 = x(s)$, $x'_1 = x(s + \Delta s)$, $x_2 = g x_1, x'_2 = g x'_1$. Let θ_1 and θ_2 denote the angles formed at x_1 and x_2 as indicated in the figure. Then upon drawing the diagonal $x_1 x'_2$ in the quadrilateral $x_1 x'_1 x'_2 x_2$, we find

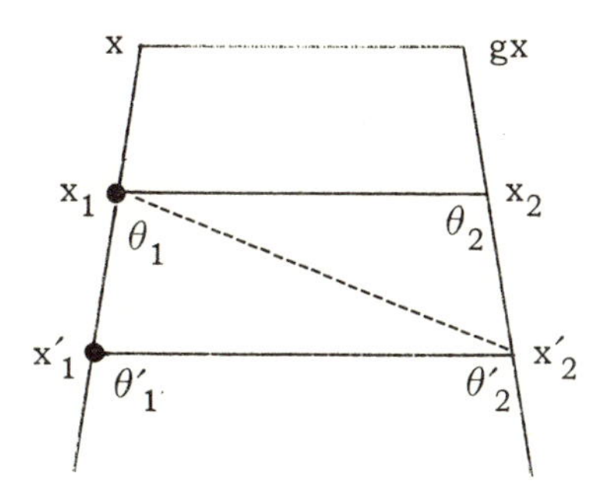

$\theta'_1 \geq \measuredangle x'_1 x_1 x'_2 + \measuredangle x_1 x'_2 x_1^2$ since the sum of the three angles in triangle $x'_1 x_1 x'_2$ does not exceed $180°$. A similar inequality holds for $\pi - \theta_2$ and adding, we get $\theta'_1 + (\pi - \theta_2) \geq \theta_1 + (\pi - \theta'_2)$. Hence

$$(\theta'_1 - \theta_1) + (\theta'_2 - \theta_2) \geq 0 .$$

From this it follows at once that

$$\frac{d\theta_1}{ds} + \frac{d\theta_2}{ds} \geq 0 .$$

We have $f'(s_0) = \left.\frac{d}{ds}\right|_{s_0} d(x(s), g x(s_0)) + \left.\frac{d}{ds}\right|_{s_0} d(x(s_0), gx(s))$, and thus

$$\begin{aligned} f'(s) &= \sin(\theta_1 - 90) + \sin(\theta_2 - 90) \\ &= -\cos\theta_1 - \cos\theta_2 . \end{aligned}$$

Hence

$$f''(s) = \sin\theta_1 \frac{d\theta_1}{ds} + \sin\theta_2 \frac{d\theta_2}{ds} .$$

At a point where $f'(s_0) = 0$, we have $\cos\theta_1 = -\cos\theta_2$, that is $\theta_1 + \theta_2 = \pi$; thus $f''(s_0) = \sin\theta_1 \left(\frac{d\theta_1}{ds} + \frac{d\theta_2}{ds}\right) \geq 0$. Therefore $f(s)$ must achieve its maximum value on the segment $[x\,y]$ at one of the endpoints. That is, $f(s) < a$ for all $x(s)$ between x and y and thus $[x, y] \subset I(g, a)$. This proves (i).

(ii) By definition $I(g) = I(g; |g|)$. Let x and y be in $I(g)$. Define $x(s)$ as above. From (i) it follows that $f'(s) = 0$ for all s and thus $\theta_1 + \theta_2 = \pi$ at all points. In particular, the sum of the four angles of the quadrilateral at x, gx, gy, y is $360°$. Hence the four points lie in a flat space F by (3.8), and as a matter of fact form a parallelogram. The geodesic line passing through x and gx is stable under g and therefore lies in $I(g)$ by (5.3.3). Similarly, the line through y and gy lies in $I(g)$. Hence the join of these two lines lies in $I(g)$. In particular, the geodesic line passing through x and y lies in $I(g)$. Thus $I(g)$ is a geodesic subspace.

(iii) If g is semi-simple and $|g| = 0$, then $pol\, g = 0$ and g is conjugate to an element of $G \cap O(n, \mathbf{R})$. Without loss of generality we can assume that $g \in G \cap O(n, \mathbf{R})$. Then the orbit of the identity under $Z(g) \cap G \cap P(n, \mathbf{R})$ is clearly the fixed point set of g and this in turn coincides with $I(g)$.

If g is semi-simple and $|g| \neq 0$, then for any x, y in $I(g)$, we have seen that the line through x and gx is a parallel translate of the line through y and gy and by (5.3.3) both are stable under $pol\, g$. Therefore for any $x \in I(g)$,

$$I(g) \subset Z(pol\, g)x\ .$$

Moreover $Z(pol\, g)x$ is a geodesic subspace of X for $x \in I(g)$.

We can apply the same argument of the foregoing paragraph to the operation of $g(pol\, g)^{-1}$ on the geodesic subspace $Z(pol\, g)x$ to conclude that

$$I(g) = Z(pol\, g)x \cap Z(g(pol\, g)^{-1})x = Z(g)x\ .$$

The "only if" part of (iv) follows from (iii). The "if" part follows from (5.3.2).

LEMMA 12.2. *Let* g *be a semi-simple element in* G. *Set* $p = pol\, g$ *and* $k = gp^{-1}$. *Then*

(i) $I(g) = I(p) \cap I(k)$.

(ii) *There is a positive constant* e *such that for all* $x \in X$,

$$d(x, I(g)) \leq e(d(x, I(p)) + d(x, I(k)) \leq 2e\, d(x, I(g)) .$$

Proof. We select $x_0 \in I(g)$. Then by Lemma 12.1 (iii)

$$I(g) = Z(g)x_0,\ I(p) = Z(p)x_0,\ I(k) = Z(k)x_0 .$$

Since $p = pol\, g$, we have $Z(g) = Z(p) \cap Z(k)$ by (2.6). Consequently,

$$I(g) = I(p) \cap I(k) .$$

Let θ denote the angle (in the G-invariant metric) between I(p) and I(k), that is, the minimum angle between a geodesic line in I(p) orthogonal to I(g) at x_0 and a geodesic line in I(k) orthogonal to I(g) at x_0. We can assume $\theta > 0$ in proving (ii).

Let $x \in X$, and let $x_0 \in I(g)$ be selected so that $d(x, x_0) = d(x, I(g))$. Introducing geodesic coordinates in X at x_0, we see that in the (Euclidean) geodesic coordinates

$$d_S(x, I(g)) = d_S(x, I(p)) \csc \theta' = d_S(x, I(k)) \csc \theta''$$

where θ' and θ'' are the angles formed by the ray x_0x with I(p) and I(k). Clearly $\theta' + \theta'' \geq \theta$, since Euclidean and Riemannian angles at x_0 coincide. Hence $\sup\{\theta', \theta''\} \geq \theta/2$ and thus by Lemma 3.2

$$\begin{aligned} d(x, I(g)) &= d(x, x_0) = d_S(x, x_0) = d_S(x, I(g)) \\ &\leq (d_S(x, I(p)) + d_S(x, I(k)) \csc \theta/2 \\ &\leq e(d(x, I(p)) + d(x, I(k)) \end{aligned}$$

with $e = \csc \theta/2$. This proves the left half of (ii). The right half follows at once from $I(g) \subset I(p)$ and $I(g) \subset I(k)$.

LEMMA 12.3. *Let* $g \in G$ *and let* s *denote the semi-simple Jordan component of* g. *Then there is a function* c *such that for all* a,

$$I(g, a) \subset T_{c(a)}(I(s)) .$$

Proof. Let t denote a semi-simple element in the smallest algebraic group of $GL(n, \mathbf{R})$ containing g. Then t lies in the smallest algebraic group containing s, and therefore $Z(s) \subset Z(t)$. Consequently $I(t) \supset I(s)$.

Let $Z(s)_{\mathbf{C}}$ denote the centralizer of s in $GL(n, \mathbf{C})$, and set $C = \{z\, g\, z^{-1}\}$ where z varies over the connected component of the identity of $Z(s)_{\mathbf{C}}$. Then C is an irreducible algebraic variety. Clearly s is the semi-simple Jordan component of each $g' \in C$. Therefore t belongs to any algebraic group containing an element $g' \in C$.

Set $p = pol\, s$ and $k = s\, p^{-1}$. Then $p = pol\, g'$ for all $g \in C$; hence s and k belong to any algebraic group containing an element of C.

We identify X with a subset of $P(n, \mathbf{R})$ in the usual way (cf. (2.11)). Replacing g by a conjugate if necessary, no generality is lost in assuming that $1 \in I(s)$ since G is transitive on X.

Let $\pi : X \to I(s)$ denote the orthogonal projection of X onto $I(s)$ and set $Y = \pi^{-1}(1)$. Then Y consists of the one parameter subgroups in X orthogonal to $I(s)$. Furthermore, $X = ZY$ where Z is the connected component of the identity in $Z(s)$.

For any element $f \in G$, set $d(f) = d(1, f) = \mathrm{Tr}\, (\log f\, {}^t f)^{\frac{1}{2}}$ and for any positive real number a, set $C_a = \{g' \in C;\, d(g') \le a\}$. Then C_a is compact and bounded subset of C lies in C_a for some finite a.

Let g_{ij} denote the complex-valued function on $GL(n, \mathbf{C}) \times C$ defined by

$$g_{ij}(h, g') = (i, j) \text{ coefficient of } h\, g'\, h^{-1}$$

and similarly define the function t_{ij} by

$$t_{ij}(h, g') = (i, j) \text{ coefficient of } h\, t\, h^{-1}$$

for all $h \in GL(n, C)$. For each (i, j) with $i < j$ the set of zeros of t_{ij} on $GL(n, C)$ contains the set of common zeros of the functions $\{g_{k\ell}^2;\ 1 \le k < j,\ j \le \ell \le n\}$ because $h\, t\, h^{-1}$ keeps invariant the subspace of C^n spanned by the last $n-j+1$ standard base vectors if $h\, g\, h^{-1}$ does. By the Hilbert Nullstellensatz therefore, t_{ij} belongs to the radical of the ideal A_j of everywhere regular rational function on $GL(n, C)$ that is generated by $\{g_{k\ell}^2;\ 1 \le k < j,\ j \le \ell \le n\}$. Set

$$\tilde{g}_j = \sum g_{k\ell}^2 \ (1 \le k < j,\ j \le \ell \le n)$$

$$\tilde{t}_j = \sum t_{k\ell}^2 \ (1 \le k < j,\ j \le \ell \le n)\ .$$

Then t_j belongs to the radical of the ideal A_j and therefore there is a positive integer q such that

$$\tilde{t}_j^{\,q} = \sum a_{k\ell}\, g_{k\ell}^2 \ (1 \le k < j,\ j \le \ell \le n)$$

where each $a_{k\ell}$ is an everywhere regular rational function on $GL(n,C) \times C$. Set

$$c_j(a) = \sup\{|a_{k\ell}(h, g')|;\ h \in O(n, R),\ g' \in C_a,\ 1 \le k < j,\ j \le \ell \le n\}\ .$$

Then $c_j(a) < \infty$ since $O(n, R) \times C_a$ is compact, and

$$|\tilde{t}_j(h, g')|^q \le c_j(a) \sum g_{k\ell}^2(h, g') \ (1 \le k < j,\ j \le \ell \le n)$$

for all $(h, g') \in O(n, R) \times C_a$. Thus there is a positive integer q and a positive constant c' such that on $O(n, R) \times C_a$

$$c\, \tilde{t}_j^{\,q} \le \tilde{g}_j\ . \qquad (12.3.1)$$

Given $x \in P(n, R)$, there is an $h_x \in O(n, R)$ such that $h \times h^{-1} = \operatorname{diag}(\lambda_1, \ldots, \lambda_n)$ with

$$\lambda_1 \le \lambda_2 \le \cdots \le \lambda_n$$

by the principal axis theorem. Therefore

$$d(x, g'x) = \left(\mathrm{Tr}\left(\log^2 x^{-\frac{1}{2}} g'x\, {}^t g'x^{-\frac{1}{2}}\right)\right)^{\frac{1}{2}}$$

$$\geq \log n^{-1}\ \mathrm{Tr}\ x^{-\frac{1}{2}} g'x\, {}^t g'x^{-\frac{1}{2}}, \text{ by Lemma 7.6 (i)}$$

$$\geq \log n^{-1}\ \mathrm{Tr}\ x^{-1} g'x\, {}^t g' .$$

We have for all $h \in O(n, R)$, $g' \in C$

$$\mathrm{Tr}\ x^{-1} g'x\, {}^t g' = \mathrm{Tr}\,(h x h^{-1})^{-1} h g' h^{-1} (h x h^{-1} ({}^t(h g' h^{-1})$$

and therefore

$$\mathrm{Tr}\ x^{-1}\ g' x\, {}^t g' = \sum_{1 \leq i, j \leq n} g_{ij}^2 \lambda_i^{-1} \lambda_j$$

where we have written merely g_{ij} for $g_{ij}(h_x, g')$.

Now

$$\sum_{j=2}^{n} \tilde{g}_j \lambda_{j-1}^{-1} \lambda_j = \sum_{j=2}^{n} \sum_{k,\ell} g_{k\ell}^2 \lambda_{j-1}^{-1} \lambda_j \ (1 \leq k < j,\ j \leq \ell \leq n)$$

$$\leq \sum_{j=2}^{n} \sum_{k,\ell} g_{k\ell}^2 \lambda_k^{-1} \lambda_\ell \ (1 \leq k < j,\ j \leq \ell \leq n) .$$

Thus

$$\sum_{2}^{n} \tilde{g}_j \lambda_{j-1}^{-1} \lambda_j \leq n \sum_{1 \leq i < j \leq n} g_{ij}^2 \lambda_i^{-1} \lambda_j . \tag{12.3.2}$$

On the other hand,

$$\sum_{2}^{n} \tilde{g}_j \lambda_{j-1}^{-1} \lambda_j = \sum_{1 \leq i < j \leq n} g_{ij}^2 (\lambda_i^{-1} \lambda_{i+1} + \dots + \lambda_{j-1}^{-1} \lambda_j) .$$

Inasmuch as $\sum_{1}^{k} x_i \geq k(x_1 x_2 \dots x_k)^{\frac{1}{k}}$ for any positive $x_1, \dots, x_k$ we see that

$$\sum_{2}^{n} \tilde{g}_j \lambda_{j-1}^{-1} \lambda_j \geq \sum_{1 \leq i < j \leq n} g_{ij}^2 (\lambda_i^{-1} \lambda_j)^{\frac{1}{n}} . \tag{12.3.3}$$

We have

$$\sum_{1<i<j\le n} g_{ij}^2 \lambda_i^{-1} \lambda_j \ge n^{-1} \sum_2^n \tilde{g}_j \lambda_{j-1}^{-1} \lambda_j \quad \text{by (12.3.2)}$$

$$\ge c' n^{-1} \sum_2^n \tilde{t}_j^{\,q} \lambda_{j-1}^{-1} \lambda_j \quad \text{by (12.3.1)}$$

$$\ge c' n^{-q} \left(\sum_2^n \tilde{t}_j \lambda_{j-1}^{-\frac{1}{q}} \lambda_j^{\frac{1}{q}} \right)^q$$

since $\left(\sum_1^k x_i\right)^q \le k^{q-1} \left(\sum_1^k x_i^q\right)$ for positive x_i. By (12.3.3),

$$\sum_2^n \tilde{t}_j \lambda_{j-1}^{-\frac{1}{q}} \lambda_j^{\frac{1}{q}} \ge \sum_{1\le i<j\le n} t_{ij}^2 \left(\lambda_i^{-\frac{1}{q}} \lambda_j^{\frac{1}{q}}\right)^{\frac{1}{n}} .$$

Therefore

$$\sum_{1\le i<j\le n} g_{ij}^2 \lambda_i^{-1} \lambda_j \ge c' n^{-q} \left(\sum_{1\le i<j\le n} t_{ij}^2 \lambda_i^{-\frac{1}{nq}} \lambda_j^{\frac{1}{nq}} \right)^q .$$

Set $m = nq$. For any $x \in X$ and $h_x \in O(n, R)$

$$\sum_{i\ge j} t_{ij}^2 (h_x) \lambda_i^{-1} \lambda_j$$

is uniformly bounded since $\lambda_1 \le \lambda_2 \le \cdots \le \lambda_n$.

Therefore, there is a constant c_1 depending on a such that for all $g' \in C_a$,

$$\text{(12.3.4)} \quad d(x, g'x) \ge c_1 + q \log \operatorname{Tr} x^{-\frac{1}{m}} t\, x^{\frac{1}{m}}\, {}^t t$$

$$\ge c_1 + q\, n^{-\frac{1}{2}} \left(\operatorname{Tr} \log^2 x^{-\frac{1}{m}} t\, x^{\frac{1}{m}}\, {}^t t \right)^{\frac{1}{2}} \quad \text{by Lemma 7.6 (i)}$$

$$\ge c_1 + c_2\, d\left(x^{\frac{1}{m}}, t\, x^{\frac{1}{m}}\right) .$$

Write $g' = k \cdot p \cdot u'$ where $s = kp$ is the semi-simple Jordan component of g' and $p = pol\, g'$. Choosing in turn $t = k$ and $t = p$, we get by (12.3.4) a constant c_1 and a positive constant c_2 such that

$$(12.3.5) \qquad d(x, g'x) \geq c_1 + c_2\, d\left(x^{\frac{1}{m}}, k\, x^{\frac{1}{m}}\right)$$

$$d(x, g'x) \geq c_1 + c_2\, d\left(x^{\frac{1}{m}}, p\, x^{\frac{1}{m}}\right).$$

Inasmuch as k is a rotation about the geodesic subspace $I(k)$, we get

$$(12.3.6) \qquad d\left(x^{\frac{1}{m}}, k\, x^{\frac{1}{m}}\right) \geq c_3\, d\left(x^{\frac{1}{m}}, I(k)\right).$$

On the other hand,

$$(12.3.6') \qquad d\left(x^{\frac{1}{m}}, p\, x^{\frac{1}{m}}\right) \geq c_4\, d\left(x^{\frac{1}{m}}, I(p)\right)^{\frac{1}{2}}.$$

To prove this, we can assume that $p \neq 1$; otherwise the assertion is trivial. Let L denote the geodesic line through 1 and p; that is $L = \exp \mathbf{R} \log p$; and let L' denote the geodesic line through $x^{\frac{1}{m}}$ and $p\, x^{\frac{1}{m}}$. Let ρ denote the restriction to L' of the orthogonal projection of X onto L. Then for all $x \in X$, the Poisson bracket $[\log x, \log p] = 0$ if and only if $x \in I(p)$. Moreover, for any $x \in X$, $d(x, L) \geq d(x, I(p))$. It follows now from Lemma 6.3 that the Jacobian of the map ρ is less than some constant times $d(q, I(p))^{-\frac{1}{2}}$ for all q on the line segment $[x^{\frac{1}{m}}, p\, x^{\frac{1}{m}}]$. From this (12.3.6) follows. Let θ_k, θ_p denote the angles formed at 1 by the line $[1, x]$ and $I(k)$, $I(p)$ respectively. Then by Lemma 3.4 (ii)

$$d\left(x^{\frac{1}{m}}, I(k)\right) \geq d\left(x^{\frac{1}{m}}, 1\right) \sin \theta_k = m^{-1}\, d(x, 1) \sin \theta_k .$$

Thus

$$(12.3.7) \qquad d\left(x^{\frac{1}{m}}, I(k)\right) \geq m^{-1} \sin \theta_k\, d(x, I(k))$$

and

(12.3.7′) $$d\left(x^{\frac{1}{m}}, I(p)\right) \geq m^{-1} \sin \theta_p \, d(x, I(p)) \ .$$

We now choose the element x in the subset Y. Since $Y = \pi^{-1}(1)$, the geodesic through 1 and x is orthogonal to $I(s)$. Since $I(s) = I(k) \cap I(p)$, we find θ_k is the angle formed by the segment $[1, x]$ and a geodesic line L_k in $I(k)$ orthogonal to $I(s)$ at 1. Similarly θ_p is the angle formed by $[1, x]$ and a geodesic line L_p in $I(p)$ orthogonal to $I(s)$ at 1. Let θ denote the angle between $I(k)$ and $I(p)$. We lose no generality in assuming that $\theta \neq 0$; otherwise we would have $I(s) = I(k)$ or $I(s) = I(p)$ and Lemma 12.3 would follow at once from (12.3.6) and (12.3.7) or (12.3.6′) and (12.3.7′). Inasmuch as

$$\theta_k + \theta_p \geq \theta$$

we leave $\sup\{\theta_k, \theta_p\} \geq \theta/2$. Assume for definiteness that $\theta_p \geq \theta/2$. Then $d(s, I(s)) \sin \theta_p = d(x, 1) \sin \theta_p \leq d(x, I(p))$, and thus $d(x, I(p)) \geq \sin \theta/2 \, d(x, I(s))$. Therefore, for any $x \in Y$ and $g' \in C_a$

(12.3.8) $$d(x, g'x) \geq c_1 + c_2 \, d\left(x^{\frac{1}{m}}, p\, x^{\frac{1}{m}}\right), \text{ by (12.3.5)}$$

$$\geq c_1 + c_2\, c_4\, d\left(x^{\frac{1}{m}}, I(p)\right)^{\frac{1}{2}}, \text{ by (12.3.6′)}$$

$$\geq c_1 + m^{-1} \sin \theta_p \, c_2\, c_4\, d(x, I(p))^{\frac{1}{2}} \text{ by (12.3.7′)}$$

$$\geq c_1 + m^{-1} c_2\, c_4 \sin^2 \theta/2 \, d(x, I(s))^{\frac{1}{2}}$$

where c_2, c_4 are positive constants depending on a, and the constant c also depends on a. From (12.3.8) we infer for all $g' \in C_a$

(12.3.9) $$d(x, g'x) \geq c_5(a)\, d(x, I(s))^{\frac{1}{2}}$$

for all $x \in Y$ such that $d(x, I(s))$ is sufficiently large where $c_5(a)$ is a positive constant depending on a.

We have seen that any point in X can be expressed as $z \cdot x$ with $z \in Z$ and $x \in Y$. For any $z \in Z$ and $x \in Y$,

$$(12.3.10) \qquad \begin{aligned} d(zx, g\,zx) &= d(x, z^{-1} g\,zx) \geq d(\pi(x), g'\,\pi(x)) \\ &\geq d(1, g'1) = d(g') \end{aligned}$$

where $g' = z^{-1} g z$, since the orthogonal projection $\pi : X \to I(s)$ diminishes distances. From (12.3.10) we infer that if $a \geq d(zx, g\,zx)$, then $a \geq d(g')$ and

$$(12.3.11) \qquad d(zx, g\,zx) \geq c_5(a)\, d(x, I(s))^{\frac{1}{2}} .$$

Since I(s) is stable under each $z \in Z$, we have $d(zx, I(s)) = d(x, I(s))$ and consequently for all $x \in X$

$$d(x, I(s))^{\frac{1}{2}} \leq \frac{a}{c_5(a)}$$

if $d(x, g\,x) \leq a$. Set $c(a) = (a/c_5(a))^2$. Then for all $x \in I(g, a)$

$$d(x, I(s)) < c(a) .$$

Therefore $I(g, a) \subset T_{c(a)}\, I(s)$.

LEMMA 12.4. *Let* G *be a semi-simple analytic linear group, let* X *be its associated symmetric space, let* $g \in G$, *and let* s *be the semi-simple Jordan component of* G. *Then for any real number* a,

$$hd(I(g, a), I(s)) = \infty, \quad \textit{if } g \neq s .$$

In any case $I(g, a) \cap I(s)$ *contains balls in* I(s) *of arbitrarily large radius.*

Proof. Let $u = s^{-1}g$ and suppose $u \neq 1$. Then $u \in Z(s)$ the centralizer of s in G. Since Z(s) is a reductive group (cf. 2.6), we can embed u in a three dimensional subgroup of $Z(s)$ whose Lie algebra has generators h, n, n_- with $\exp n = u$ and

$$[h, n] = 2n, [h, n_-] = -2n_-, [n, n_-] = h$$

(cf. 2.9). As a result for any $x \in I(s)$

$$\begin{aligned} d((\exp -th)x, g(\exp -th)x &= d(s^{-1} \cdot x, (\exp th)u(\exp -th)x) \\ &\geq d((\exp th)u(\exp -th)x, x) - |s| \end{aligned}$$

where $|s| = d(sx, x)$. Now $(\exp th)u(\exp -th) = \exp n(t)$ where $n(t) = e^{2t}n$. Hence $d((\exp n(t)x, x) \to \infty$ as $t \to \infty$ and consequently $(\exp -th)x$ does not belong to $I(g, a)$ for $a < d((\exp n(t))x, x) - |s|$. Inasmuch as $T_b(I(g, a)) \subset I(g, a + 2b)$ for any positive a and b, it follows that

$$d((\exp -th)x, I(g, a)) \to \infty \quad \text{as} \quad t \to \infty .$$

From this the first assertion of the lemma follows. The second assertion comes from considering balls around $(\exp th)x$ and using the fact that $(\exp -th)u(\exp th) \to 1$ as $t \to \infty$.

We can now give a criterion for semi-simplicity and unipotence of elements of G that is preserved by pseudo-isometric Γ-morphisms of X.

PROPOSITION 12.5. *Let* G *be a semi-simple analytic linear group, let* X *be the associated symmetric space, let* $g \in G$ *and let* $I(g, a) = \{x \in X; d(x, gx) \leq a\}$. *Then*

(i) g *is semi-simple if and only if for every* a, $I(g, a)$ *lies within finite Hausdorff distance of a geodesic subspace of* X;

(ii) *Assume that* G *has no center. Then* g *is unipotent if and only if for every* $a > 0$, $I(g, a)$ *contains some arbitrarily large balls of* X.

Proof. Let I be a geodesic subspace such that $hd(I(g, a), I) < \infty$. Then $I \subset T_c(I(g, a))$ and $I(g, a) \subset T_c(I)$ for some finite c. Clearly $g\, I(g, a) = I(g, a)$, for $d(x, gx) = d(gx, g \cdot gx)$. Hence $g^n\, I(g, a) = I(g, a)$ for every n, and $g^n I \subset T_c(g_n I(g,a) \subset T_c(I(g,a))$. By Lemma 5.4, gI is a parallel translate of a geodesic subspace of I, and hence gI is a parallel translate of I – for dimension reasons. Thus $gI = hI$ with $h \in Z(pol\, I)$, where $pol\, I$ denotes the subgroup of the stabilizer G_I generated by the polar subgroups of G_I. We have $g^n I \subset T_c(I(g,a)) \subset T_{2c}(I)$. In particular, $hI = gI \subset T_{2c}(I)$ and we can select $h \in Z(pol\, I)$ so that $gI = hI$ and $d(x, hx) = d(I, hI) \leq 2c$ for all $x \in I$. It follows that for all $x \in I$, $d(x, h^{-1} gx) \leq d(x, gx) + d(gx, h^{-1} gx) \leq a + 4c$. By Lemma 5.4, $h^{-1} g$ moves each geodesic line in I parallel to itself and therefore $h^{-1} g \in Z(pol\, I)$. Consequently, $g = h \cdot h^{-1} g \in (pol\, I)$.

We can assume without loss of generality that G is a self-adjoint group and that I contains a point x_0 fixed under $G \cap O(n, R)$. Then G_I and $Z(pol\, I)$ are self-adjoint groups and we have by (5.4.1)

$$I = (pol\, I \cap P(n,R)) \cdot (G_I \cap O(n,R))$$

$$pol\, I = (pol\, I \cap P(n,R)) \cdot (pol\, I \cap O(n,R)) .$$

Set $Z = Z(pol\, I)$. Since $Z(pol\, I)$ is self-adjoint, we have

$$Z = (Z \cap P(n,R)) \cdot (Z \cap O(n,R)) .$$

For any $p \in pol\, I$ and $z \in Z \cap O(n,R)$ we have $z p x_0 = p z x_0 = p x_0$. Since $I = pol\, I\, x_0$, we see that $Z \cap O(n,R) \subset G_I$, in fact $Z \cap O(n,R)$ fixed each point in I.

The orbit $Z x_0$ is a geodesic subspace by (3.4.1) and $pol\, Z x_0$ commutes elementwise with $pol\, I$ since $pol\, Z x_0 \subset Z$. It follows by Lemma 5.5 that the geodesic lines in $Z x_0$ orthogonal to I form a geodesic subspace D. Clearly D is a set of representatives for the parallel translates of I, and Z operates on D via $y \to z(pol\, I)y \cap D$ for $y \in D$, $z \in Z$. The convex hull of $\{g^n I;\ n = 0, \pm 1, \ldots\}$ intersects D in a bounded convex set since $g^n I \subset T_{2c}(I)$ for each n. Therefore in its action on D, g keeps invariant a closed bounded convex set. It follows (from the Brouwer fixed point theorem) that g keeps fixed a point in D under the above action; that is, g stabilizes a parallel translate of I under the canonical action. Therefore there is an element $z \in Z$ such that $gzI = zI$; i.e., $z^{-1}gz \in G_I \cap Z$. We have $G_I \cap Z = (G_I \cap Z \cap P(n,R))(G_I \cap Z \cap (n,R))$. From the expression for G_I, we see that each element in $G_I \cap Z \cap P(n,R)$ commutes with each element in $G_I \cap Z \cap O(n,R)$. Consequently, every element in $G_I \cap Z$ is semi-simple and in particular, g is semi-simple.

We have thus proved: If $I(g,a)$ lies within finite Hausdorff distance of a geodesic subspace, then g is semi-simple. The converse follows immediately from Lemma 12.3 which implies that $hd(I(g,a), I(g)) < \infty$ if g is a semi-simple element.

If g is a unipotent element, then its semi-simple part is the identity element and therefore $I(g,a)$ contains arbitrarily large balls in X by Lemma 12.4. This proves the "only if" assertion of (ii). Conversely, suppose that $I(g,a)$ contains arbitrarily large balls of X. Let s denote the semi-simple part of g. Then arbitrarily large balls of X lie within $T_c(I(s))$ for some finite c. It follows at once $I(s) = X$. This implies that s is central in G. By our hypothesis on G, $s = 1$; that is, g is unipotent.

PROPOSITION 12.6. *Let* G *and* G′ *be semi-simple analytic linear groups, let* Γ *and* Γ' *be subgroups of* G *and* G′, *and let* X *and* X′ *denote the symmetric Riemannian spaces associated to* G *and* G′ *respectively. Let* $\theta : \Gamma \to \Gamma'$ *be an isomorphism, let* $\phi : X \to X'$ *and* $\phi' : X' \to X$ *be pseudo-isometries equivariant with respect to* θ *and* θ^{-1} *respectively. Then*

(i) θ *sends semi-simple elements to semi-simple elements and vice-versa*

(ii) *if* G′ *has no center, then* θ *sends unipotent elements to unipotent elements.*

Proof. Let $g \neq e$ be an element in Γ, let $g = su$ be its Jordan decomposition into semi-simple part s and unipotent part u, set $g' = \theta(g)$, and let $s'u'$ be the Jordan decomposition of g'. Assume that ϕ and ϕ' are (k, b) pseudo-isometries.

We begin with the simple observations that

$$T_c(I(g, a)) \subset I(g, a + 2c)$$

$$\phi(I(g, a)) \subset I(g', ka)$$

for any c and a. By Lemma 10.1′, $\dim X = \dim X'$ and both ϕ and ϕ' are surjective. It follows that for $a > b$,

$$\phi(I(g, a)) \supset I(g', k^{-1} a) .$$

Set $m = \dim I(s)$, $n = \dim I(s')$. Let π and π' denote the orthogonal projections of X onto $I(s)$ and of X' onto $I(s')$ respectively. Let ψ denote the restriction to $I(s)$ of $\pi' \circ \phi$.

To prove (i), we assume that g is semi-simple, i.e., $g = s$. Then $\psi(I(g)) = \phi(I(g, |g|) \subset I(g', k|g|) \subset T_c(I(s'))$ for some finite c, by Lemma 12.3. Consequently, for any $x \in I(s)$ and $y \in I(s)$,

$$\begin{aligned} d(\psi(x), \psi(y)) &\geq d(\phi(x), \phi(y)) - 2c \\ &\geq k^{-1}(d - 2kc), \; d = d(x, y) \\ &\geq 2^{-1} k^{-1} d \end{aligned}$$

for $d > 4kc$. Thus ψ is a $(2k, 4kc)$ pseudo-isometry of $I(s)$ into $I(s')$. It follows by Lemma 10.1′ that $n \geq m$ since $I(s)$ and $I(s')$ are spaces of non-positive curvature.

Let ψ' denote the restriction of $\pi \circ \phi'$ to $I(s')$. By the observations above and Lemma 12.3

$$\phi'(I(g', a)) \subset I(g, ka) \subset T_c(I(s))$$

for some finite c. Therefore ψ' is a pseudo-isometry on $I(g', a) \cap I(s')$. This latter set contains arbitrarily large balls of $I(s')$ by Lemma 12.4.

Applying Lemma 10.1′, we find $\dim I(s) \geq \dim I(s')$. From $n \geq m$ and $m \geq n$, we conclude $m = n$. Since $\psi : I(s) \to I(s')$ is a pseudo-isometry, it is surjective by Lemma 10.1′. Consequently, $I(s') = \psi(I(g)) = \pi'(\phi(I(g))$ $\subset T_c(\phi(I(g))$, since $\phi(I(g)) \subset T_c(I(s'))$ as stated above, and thus

$$I(s') \subset T_c(I(g', k|g|) .$$

That $I(g', a) \subset T_{c(a)}(I(s'))$ for all a is asserted in Lemma 12.3. Thus for each $a > |g'|$, we have

$$hd(I(g', a), I(s')) < \infty .$$

By Lemma 12.4, g' is semi-simple. Interchanging the role of θ and θ^{-1} we see that θ^{-1} sends semi-simple elements to semi-simple elements. This completes the proof of (i).

Proof of (ii). Suppose that g is unipotent. By Lemma 12.4, $I(g, a)$ contains some arbitrarily large balls of X. By Lemma 10.1′, it follows that $I(g', ka)$, which contains $\phi(I(g, a))$ has the same property. If G' has no center, g' is unipotent by Proposition 12.5.

§13. The Basic Approximation

Throughout this section, we fix the following notation and assumptions.

Let G be a semi-simple analytic group having a faithful matrix representation, let K be a maximal compact subgroup, and let X denote the symmetric Riemannian space G/K. Let G', K', X' be a similar triple. Let Γ and Γ' be discrete subgroups of G and G' respectively such that G/Γ and G'/Γ' have finite measure. Finally, let $\theta : \Gamma \to \Gamma'$ be an isomorphism. Assume there exist maps $\phi : X \to X'$ and $\phi' : X' \to X$ which are (k, b) pseudo-isometries equivariant with respect to θ and θ^{-1} respectively. We note that ϕ is surjective by Lemma 10.1′.

LEMMA 13.1. *Let* F *denote a* Γ*-compact* r*-flat in* X, *let* Δ *denote the stabilizer of* F *in* Γ, *and set* $\Delta' = \theta(\Delta)$. *Then there is a unique* r*-flat* F' *in* X' *stable under* Δ', F' *is* Γ'*-compact, and* Δ' *is the stabilizer of* F' *in* Γ'.

Proof. Let G_F denote the stabilizer of F in G and set $A = pol\, F$. Then by Lemma 5.1, A is a maximal polar subgroup of G, A operates simply transitively on F, and G is the normalizer of A in G.

We claim that every element in G_F is semi-simple. For let $g \in G_F$ and let $g = su$ be the Jordan product decomposition with s semi-simple and u unipotent. By 2.1, $\{\exp t \log u;\ t \in \mathbf{R}\}$ lies in any algebraic group containing u and therefore lies in the connected component of the identity of G_F. Since A is a maximal polar subgroup of G, we get by (2.3.iv) that the centralizer of A is of finite index in the normalizer of A. Hence $u \in Z(A)$. By (2.6.iv), $Z(A) = M \times A$ (direct) with M compact. It follows at once that every element of $Z(A)$ is semi-simple and in particular $u = 1$. Thus $g = s$ and every element is G_F is semi-simple.

Since $\Delta \cap Z(A)$ is of finite index in Δ and $\Delta \cap M$ is finite, we can select a subgroup of Δ_1 of finite index in Δ such that $\Delta_1 \subset Z(A)$ and $\Delta_1 \cap M = (1)$. Then Δ_1 is an abelian subgroup since it injects into the abelian group $M \backslash Z(A)$.

By hypothesis F is Γ-compact. Hence $\Gamma \backslash \Gamma F$ is compact. From $\Gamma \backslash \Gamma F = \Delta \backslash F$ and $\Delta_1 \backslash \Delta$ finite we get that $\Delta_1 \backslash F$ is compact. Since A operates simply transitive on F, we have $F = M \backslash MA$ and hence

$$\Delta_1 \backslash F = M\Delta_1 \backslash MA = M\Delta_1 \cap A \backslash A .$$

Therefore $M\Delta_1 \cap A \backslash M$ is compact. Since $\Delta_1 = M \cap \Delta_1 \backslash \Delta_1 = M \backslash M\Delta_1 = M \backslash M\Delta_1 = M\Delta_1 \cap A$, we get $\operatorname{rank} \Delta_1 = \dim A = \mathbb{R}\text{-rank } G = r$.

By Proposition 12.6, θ sends semi-simple elements of Γ to semi-simple elements of Γ' and vice-versa. Therefore, by Lemma (11.3.iii)$'$, $\mathbb{R}$-rank $G = \mathbb{R}$-rank G'. Thus $\theta(\Delta_1)$ is an abelian subgroup of semi-simple elements of Γ' of rank equal to $\mathbb{R}$-rank G' and thus contains an $\mathbb{R}$-hyperregular element γ' of G'. By Raghunathan's criterion $Z(\gamma')/Z(\gamma') \cap \Gamma'$ is compact. Moreover, the element γ' is polar-regular since any $\mathbb{R}$-hyperregular is polar-regular.

By Lemma 5.2 (ii), γ' stabilizes a unique r-flat F' in X'. By Lemma 5.2 (iii), $Z(\gamma')$ stabilizes F' and acts transitively on F'. Hence $Z(\gamma') \cap \Gamma' \backslash F'$ is a quotient of $Z(\gamma') \cap \Gamma' \backslash Z(\gamma')$ and is compact; that is, F' is a Γ'-compact r-flat of X'.

Let $\Gamma'_{F'}$ denote the stabilizer of F' in Γ'. Applying to $\Gamma'_{F'}$ the result proved above for Δ, we see that $\Gamma'_{F'}$ contains an abelian subgroup of finite index of rank r. Since $\theta(\Delta_1)$ is an abelian subgroup of $\Gamma'_{F'}$ of rank r, we see that $\theta(\Delta_1)$ is of finite index in $\Gamma'_{F'}$. Since Δ_1 is of finite index in Δ, so also is the subgroup $\bigcap_{g \in \Delta} g\Delta_1 g^{-1}$. Thus without loss of generality we may assume that Δ_1 is normal in Δ and $\theta(\Delta_1)$ is normal in $\theta(\Delta)$. For any $g' \in \theta(\Delta)$, $\theta(\Delta_1)g'F' = g'\theta(\Delta_1)F' = g'F'$;

that is, $\theta(\Delta_1)$ stabilizes the r-flat $g'F'$. Since $\theta(\Delta_1)$ stabilizes a unique r-flat, we get $g'F' = F'$ for all $g' \in \theta(\Delta)$. Thus $\theta(\Delta) \subset \Gamma'_{F'}$. Applying the same result to θ^{-1}, we get $\theta^{-1}(\Gamma'_{F'}) \subset \Delta$. Thus $\theta(\Delta) = \Gamma'_{F'}$.

REMARK. In the case that G/Γ and G'/Γ' are compact, there is no need above to employ Raghunathan's criterion, since in this case, all the elements of Γ and Γ' are semi-simple (cf. Section 11).

LEMMA 13.2. *Let* F *be a* Γ*-compact* r*-flat in* F *and let* F' *denote the unique* r*-flat in* F' *stabilized by* $\theta(\Gamma_F)$. *Then there is a constant* v *depending only on* k *and* b *but not on the particular choice of the* Γ*-compact* r*-flat* F *such that*

$$\mathrm{hd}(\phi(F), F') \le v .$$

Proof. Let $\pi: X \to F$ and $\pi': X' \to F'$ denote the orthogonal projections of X onto F and of X' onto F'. We note that $\pi(\gamma x) = \gamma\pi(x)$ for all $\gamma \in \Gamma_F$ and $x \in X$, and $\pi'(\gamma' x) = \gamma'\pi'(x)$ for all $\gamma' \in \Gamma'_{F'}$ and $x \in X'$.

We claim first that $\pi'(\phi(F)) = F'$. We can deduce this from the results of Section 12, using Lemma 12.3 to show that $\pi' \circ \phi$ is a pseudo-isometry of F onto F'. Instead, however, we present here a simple topological argument.

Let Δ be a free abelian subgroup of finite index in Γ_F, and let ψ denote the restriction of $\pi' \circ \phi$ to F. Then ψ may be regarded as a Δ-bundle map of F to F'. Set $\Delta' = \theta(\Delta)$ and let ψ_Δ denote the induced map of $\Delta\backslash F$ to $\Delta'\backslash F'$. Since the stabilizer G_x of any point x in X is compact, the subgroup $\Delta \cap G_x$ is finite for all $x \in X$ and thus contains only the identity element. Hence Δ operates freely on F and F is a principal Δ-bundle. Furthermore, since F is a contractible space, F is a universal Δ-bundle. Similarly F' is a universal Δ'-bundle. As is well-known from the elementary theory of universal bundles, the map $\psi_\Delta : \Delta\backslash F \to \Delta'\backslash F'$ is a homotopy equivalence. Inasmuch as $\Delta\backslash F$ is an r-dimensional torus, ψ_Δ induces an isomorphism of the homology group $H_r(\Delta\backslash F)$ onto $H_r(\Delta'\backslash F')$. Inasmuch as $\Delta'\backslash F'$ is a compact manifold, any top dimensional cycle has the entire space $\Delta'\backslash F'$ as its support. Hence $\psi_\Delta(\Delta\backslash F) = \Delta'\backslash F'$. Consequently $\psi(F) = F'$ and $\pi'(\phi(F)) = F'$.

Next we claim that $\pi(T_b(\phi^{-1}(F'))) = F$. We can assume $k \geq 1$ and $b > 0$. Let $\phi_\Delta : \Delta\backslash X \to \Delta'\backslash X'$, $\pi_\Delta : \Delta\backslash X \to \Delta\backslash F$, and $\pi_{\Delta'} : \Delta'\backslash X' \to \Delta'\backslash F'$ be the maps induced by ϕ, π, and π'. We can assume, upon replacing Δ by a subgroup of finite index if necessary that the canonical projections of X onto $\Delta\backslash X$ and F' onto $\Delta'\backslash F'$ are injective on all balls in X of radius $2b$ and all balls in F' of radius $6kb$ (cf. Lemma 5.3). Then balls in $\Delta\backslash X$ of radius b and in $\Delta'\backslash F'$ of radius $3kb$ are convex. Let Σ be a triangulation of the r-dimensional torus $\Delta'\backslash F'$ having mesh less than b/k, and let Σ_0 denote the set of vertices in Σ. For each vertex $p \in \Sigma_0$, select $\xi(p)$ to be any point in $\phi_\Delta^{-1}(p)$. One can extend ξ to a map of $\Delta'\backslash F'$ into $\Delta\backslash X$ so that $\phi_\Delta(\xi(\sigma)) \subset T_{bk}(\sigma)$ for each simplex σ by induction on the dimension of σ: for each k-simplex σ in Σ, select a vertex p_σ and map each shortest geodesic segment $[p_\sigma, p]$ to the shortest geodesic segment $[\xi(p_\sigma), \xi(p)]$ where p varies over the face of σ opposite to p_σ. For each $p \in \Delta'\backslash F'$, $\phi_\Delta(\xi(p))$ lies in a ball of center p and radius $3bk$ and thus $\phi_\Delta \circ \xi$ can be deformed along shortest segments into the identity map. Consequently, ξ sends a fundamental cycle in $H_r(\Delta'\backslash F')$ to a fundamental cycle of $H_r(\Delta\backslash X)$.

Now the spaces F and X are both universal Δ-bundles and therefore the inclusion map $F \to X$ induces a homotopy equivalence $\Delta\backslash F \to \Delta\backslash X$. In particular $H_r(\Delta\backslash F) \approx H_r(\Delta\backslash X)$ and the map $\pi_\Delta : \Delta\backslash X \to \Delta\backslash F$ induces an isomorphism of $H_r(\Delta\backslash X)$ onto $H_r(\Delta\backslash F)$. Consequently $\pi_\Delta \circ \xi$ sends a fundamental cycle in $H_r(\Delta'\backslash F')$ to a fundamental cycle h in $H_r(\Delta\backslash F)$, and $\pi_\Delta \circ \xi(\Delta'\backslash F') = \Delta\backslash F$, since $\Delta\backslash F$ is a compact r-dimensional manifold and is the support of the cycle h. Clearly

$$\xi(\Delta'\backslash F') \subset T_b(\phi_\Delta^{-1}(\Delta'\backslash F')) .$$

Consequently

$$\Delta\backslash F = \pi_\Delta(T_b(\phi_\Delta^{-1}(\Delta'\backslash F')) .$$

It follows at once that

$$F = \pi(T_b(\phi^{-1}(F'))$$

since both sides are stable under Δ.

Set

$$d = \sup_{x \in F} d(\phi(x), F') \tag{13.2.1}$$

$$d' = \sup_{x \in F'} d(\phi(F), x) . \tag{13.2.2}$$

From $d(\phi(x), F') = d(\phi(x), \pi'(\phi(x))$ and $\pi'(\phi(F)) = F'$, we see that

$$F' \subset T_d^*(\phi(F))$$

where T_d^* denote the *closed* tube of radius d. It follows, by definition of d' that

$$d' \leq d .$$

Applying ϕ^{-1}, one gets from (13.2.2)

$$\sup_{x \in \phi^{-1}(F')} d(F, x) \leq \sup\{kd', b\} .$$

Since $d(F, x) = d(\pi(x), x)$ for any $x \in X$, we get

$$\sup_{x \in \phi^{-1}(F')} d(\pi x, x) \leq \sup\{kd', b\} .$$

Inasmuch as $\pi(T_b(\phi^{-1}(F')) = F$, we infer

$$F \subset T^*_{kd'+b}(\phi^{-1}(F')), \text{ if } b \leq kd'$$

and

$$F \subset T^*_{2b}(\phi^{-1}(F')), \text{ if } b \geq kd' .$$

Hence, applying ϕ

$$\phi(F) \subset T^*_{k(kd'+b)}(F'), \text{ if } b \leq kd'$$

and

$$\phi(F) \subset T^*_{2kb}(F'), \text{ if } b \geq kd' .$$

Thus, if $b \leq kd'$, $d \leq k^2 d' + kb \leq 2k^2 d'$. That is,

$$d' \leq d \leq 2k^2 d', \text{ if } b \leq kd' \tag{13.2.3}$$

and

$$d' \leq d \leq 2kb \quad , \text{ if } b \geq kd' .$$

It remains to consider the case $b \leq kd'$. Set

$$\begin{aligned} h(x) &= d(x, F') && \text{for} \quad x \in \phi(F) \\ h'(x) &= d(\phi(F), x) && \text{for} \quad x \in F' \\ E_s &= X' - T_s(F'), \; s > 0 . \end{aligned}$$

The function h' defines a continuous function on $\Delta' \setminus F'$ and therefore attains its supremum. Now choose a point x_0 in F' so that $h'(x_0) = d'$. For any $s > 0$, let B'_s denote the ball in X' of radius s with center x_0. Clearly for any y in $B'_{d'/2} \cap F'$, $h'(y) \geq d'/2$. Hence if $x \in \phi(F)$ and $\pi'(x) \in B'_{d'/2} \cap F'$, then

$$d(\pi'(x), x) \geq d(\pi'(x), (F)) \geq d'/2 \; ; \tag{13.2.4}$$

that is, $h(x) \geq d'/2$. On the other hand, for all $x \in \phi(F)$, $d(\pi'(x), x) \leq d \leq 2k^2 d'$ by (13.2.1) and (13.2.3). Thus if $x \in \phi(F)$ and $\pi'(x) \in B'_{d'/2} \cap F'$, then $d(x, x_0) \leq d(x, \pi'(x)) + d(\pi'(x), x_0) \leq 2k^2 d' + d'/2 = (2k^2 + \frac{1}{2})d'$. Set $e = (2k^2 + \frac{1}{2})d'$. We can restate our last assertion as: Any point x in $\phi(F)$ for which $\pi'(x) \in B'_{d'/2} \cap F'$ must lie in $B'_e \cap \phi(F)$. Coupling this remark with (13.2.4), we get

$$\pi'(B'_e \cap E_{d'/2} \cap \phi(F)) \supset B'_{d'/2} \cap F' .$$

By Lemma 6.4, the projection π' shrinks the r-dimensional volume of any set in E_s by a factor $c\,s^{-\frac{1}{2}}$, where c is a constant depending only on the space X'. Therefore

$$m_r(B'_e \cap E_{d'/2} \cap \phi(F)) \geq c^{-1}(d'/2)^{\frac{1}{2}}\, m_r(B'_{d'/2} \cap F') \tag{13.2.5}$$

where m_r denotes r-dimensional measure.

On the other hand, for any subset $S \subset F$, $m_r(\phi(S)) \leq k^r\, m_r(S)$ since $d(\phi(x), d(y)) \leq k\, d(x,y)$ for all $x, y \in X$. Thus

$$\begin{aligned} m_r(B'_e \cap \phi(F)) &\leq k^r\, m_r(\phi^{-1}(B'_e) \cap F) \\ &\leq k^r\, m_r(B_{ke} \cap F) \end{aligned} \tag{13.2.6}$$

where B_{ke} is a ball of radius ke and center in $\phi^{-1}(x_0)$. We have for any $a > 0$, $m_r(B_{a\,d'}) = a^r\, m_r(B_{d'})$. Combining (13.2.5) and (13.2.6), we get

$$c^{-1}(d'/2)^{\frac{1}{2}}\, m_r(B'_{d'/2} \cap F') \leq k^r\, m_r(B_{ke})$$

$$c^{-1}(1/2)^{r+\left(\frac{1}{2}\right)}\, (d')^{r+\left(\frac{1}{2}\right)} \leq k^r(2k^3 + k/2)^r\, (d')^r\ .$$

Consequently

$$d' \leq c^2(2k^4 + k^2/2)^{2r}\, 2^{2r+1}\ .$$

Set

$$v = \sup\{c^2\, 2^{6r+2}\, k^{8r+2},\, 2kb\}\ . \tag{13.2.7}$$

Then $d' \leq d \leq v$, and from (13.2.1) and (13.2.2) we get $\phi(F) \subset T^*_v(F')$ and $F' \subset T^*_v(F)$. Consequently, $hd(\phi(F), F') \leq v$.

§14. The Map $\overline{\phi}$

Let $\mathfrak{F}$ and $\mathfrak{F}'$ denote the space of r-flats in X and X′ respectively, each topologized by uniform convergence on compact sets. We prove in this section that if G/Γ is compact, then ϕ induces a homeomorphism $\overline{\phi}$ of $\mathfrak{F}$ onto $\mathfrak{F}'$.

LEMMA 14.1. *Under the hypotheses of Section 13, for each* $F \in \mathfrak{F}$ *there is a unique* $F' \in \mathfrak{F}$ *such that*

$$hd(\phi(F), F') \leq v$$

where the constant v *is given by* (13.2.7). *The map* $\overline{\phi} : F \to F'$ *is a homeomorphism of* $\mathfrak{F}$ *onto* $\mathfrak{F}'$. *For all* $\gamma \in X$ *and* $F \in \mathfrak{F}$, $\overline{\phi}(\gamma F) = \theta(\gamma)\,\overline{\phi}(F)$.

Proof. As a point set, we have

$$\mathfrak{F} = G/G_F$$

since G operates transitively on the set of r-flats of X. Moreover, it is clear that the map $g\,G_F \to gF$ is a continuous map τ from G/G_F in the canonical quotient G-topology to $\mathfrak{F}$ with its topology of uniform convergence on compact sets. Since $\mathfrak{F}$ is a locally compact G-orbit and G is a countable union of compact sets, it follows by a well-known theorem that τ is open (cf. [8b] p. 7). By Lemma 8.3, the set S of Γ-compact r-flats is dense in the quotient G-topology and therefore in the given topology of $\mathfrak{F}$. For each $F_0 \in S$, let $\overline{\phi}(F_0)$ denote the unique r-flat in X′ such that

(14.1.1) $$hd(\phi(F_0), \bar{\phi}(F_0)) \le v$$

where v is the constant of Lemma 13.2. In order to prove that $\bar{\phi}$ is the restriction to S of a continuous map of $\mathfrak{F}$ into $\mathfrak{F}'$, it suffices to prove:

(14.1.2) *Given* $F \epsilon \mathfrak{F}$ *and a sequence* $\{F_n\}$ *of elements in* S *converging to* F, *then* $\{\bar{\phi}(F_n)\}$ *is convergent in* $\mathfrak{F}'$.

Proof. Let $x_0 \epsilon F$. For any $s > 0$, let B_s and B'_s denote the balls of radius s and centers x_0 in X and $\phi(x_0)$ in X' respectively. By hypothesis, $hd(g_n F \cap B_s, F \cap B_s) \to 0$ as $n \to \infty$. Since ϕ is continuous and $\phi(B_s) \supset B'_{s/k}$ for all $s \ge b$, it is seen that

(14.1.3) $$hd(\phi(F_n) \cap B'_s, \phi(F) \cap B'_s) \to 0 \text{ as } n \to \infty .$$

Let v_1 be any number such that $v_1 > v$. Let $\mathfrak{F}'_s$ denote the set of all $F'_0 \epsilon \mathfrak{F}'$ such that

$$hd(F'_0 \cap B'_s, \phi(F) \cap B'_s) \le v_1 .$$

From (14.1.1) and (14.1.3) it follows that $\bar{\phi}(F_n) \epsilon \mathfrak{F}'_s$ for all sufficiently large subscripts of the given sequence $\{F_n\}$ of Γ-compact r-flats. The set of r-flats in X' which meet the closed ball $\bar{B}'_{v_1}$ is evidently compact and contains $\mathfrak{F}'_s$ as a closed subset. The family $\{\mathfrak{F}'_s; s > 0\}$ is a nested family of compact sets and therefore has a non-empty intersection. To prove (14.1.2), it suffices to prove that $\bigcap_{s>0} \mathfrak{F}'_s$ contains only one element; for that will imply that the sequence $\{\bar{\phi}(F_n)\}$, which lies in a compact subset of $\mathfrak{F}'$, has only a single point of accumulation.

Let F' be an element in $\bigcap_{s>0} \mathfrak{F}'_s$. Then $hd(F' \cap B'_s, \phi(F) \cap B'_s) \le v_1$ for all $s > 0$. Consequently $hd(F', \phi(F)) \le v_1$ for all $v_1 > v$. Therefore

(14.1.4) $$hd(F', \phi(F)) \le v .$$

Suppose now that $F'' \epsilon \bigcap_{s>0} \mathfrak{F}'_s$. Then $hd(F'', \phi(F)) \le v$, and therefore $hd(F', F'') \le 2v$. From Lemma 7.3 (iv) it follows that $F' = F''$. This completes the proof of (14.1.2).

For each $F \in \mathfrak{F}$, set $\overline{\phi}(F) = F'$. As a restatement of (14.1.4) we have for all $F \in \mathfrak{F}$,

$$(14.1.4)' \qquad \operatorname{hd}(\overline{\phi}(F), \phi(F)) \leq v .$$

Given $\gamma \in \Gamma$ and $F \in \mathfrak{F}$, we have

$$\operatorname{hd}(\phi(\gamma F), \theta(\gamma)\phi(F)) = \operatorname{hd}(\theta(\gamma)\phi(F), \theta(\gamma)\overline{\phi}(F)) = \operatorname{hd}(\phi(F), \overline{\phi}(F)) .$$

Hence $\operatorname{hd}(\overline{\phi}(\gamma F), \theta(\gamma)\overline{\phi}(F)) \leq \operatorname{hd}(\overline{\phi}(\gamma F), \phi(\gamma F)) + \operatorname{hd}(\phi(\gamma F), \theta(\gamma)\overline{\phi}(F)) \leq 2v$. By Lemma 7.3 (iv) we see that for all $\gamma \in \Gamma$, $F \in \mathfrak{F}$

$$(14.1.5) \qquad \overline{\phi}(\gamma F) = \theta(\gamma)\overline{\phi}(F) .$$

The map $\overline{\phi}$ is injective. For given distinct elements F_0 and F in $\mathfrak{F}$, we have $\operatorname{hd}(F_0, F) = \infty$ by Lemma 7.3 (iv). Consequently $\operatorname{hd}(\phi(F_0), \phi(F)) = \infty$ since ϕ is a pseudo-isometry. Therefore $\operatorname{hd}(\overline{\phi}(F_0), \overline{\phi}(F)) = \infty$ by (14.1.4)$'$. Thus $\overline{\phi}(F_0) \neq \overline{\phi}(F)$.

To prove that $\overline{\phi}$ is in fact a homeomorphism of $\mathfrak{F}$ onto $\mathfrak{F}'$, we consider the map $\overline{\phi}' : \mathfrak{F}' \to \mathfrak{F}$ induced by the pseudo-isometric Γ-map $\phi' : X' \to X$. Let $F \in S$, and let Δ denote the stabilizer of the Γ-compact r-flat F in Γ. Set $\Delta' = \theta(\Delta)$. By Lemmas 13.1, 13.2 $\overline{\phi}(F)$ is the unique r-flat in X' stable under Δ', and $\overline{\phi}'\overline{\phi}(F)$ is the unique r-flat in X stable under Δ. Therefore $\overline{\phi}'\overline{\phi}(F) = F$ for all $F \in S$. Since S is dense in $\mathfrak{F}$, $\overline{\phi}'\overline{\phi} =$ identity. Similarly $\overline{\phi}\,\overline{\phi}' =$ identity. Therefore $\overline{\phi}$ is a homeomorphism of $\mathfrak{F}$ onto $\mathfrak{F}'$. The proof of Lemma 14.1 is now complete.

THEOREM 14.2. *Let G and G' be semi-simple analytic linear groups. Let Γ and Γ' be discrete torsion free subgroups of G and G' respectively such that G/Γ and G'/Γ' are compact. Let $\theta : \Gamma \to \Gamma'$ be an isomorphism of Γ onto Γ'. Let X and X' be the symmetric Riemannian space $X = G/K, X' = G'/K'$ where K and K' are maximal compact subgroups of G and G' respectively. Then*

(i) *there is a pseudo-isometry $\phi : X \to X'$ equivariant with respect to θ.*

(ii) *Let* $\mathfrak{F}$ *and* $\mathfrak{F}'$ *denote the space of maximal flat subspaces in* X *and* X′ *respectively. Then there is a homeomorphism* $\overline{\phi}:\mathfrak{F}\to\mathfrak{F}'$ *equivariant with respect to* θ *and a constant* v *such that for all* $F \in \mathfrak{F}'$

$$hd(\phi(F), \overline{\phi}(F)) \leq v .$$

Proof. Assertion (i) has already been proved in Lemma 9.2. Applying (i) to the isomorphism $\theta^{-1}:\Gamma'\to\Gamma$ we find the existence of a pseudo-isometry $\phi':X'\to X$ equivariant with respect to θ^{-1}. The hypotheses of Section 13 are therefore satisfied and we can apply Lemma 14.1 to obtain the desired condition.

§15. The Boundary Map ϕ_0

We have seen in Section 4 that the Furstenberg maximal boundary can be defined in terms of chambers. Thanks to the existence theorem of the preceding section, we shall prove in this section that the Γ-space pseudo-isometry $\phi: X \to X'$ induces a homeomorphism $\phi_0 : X_0 \to X'_0$ of the maximal boundaries. To accomplish this, we must show that for any chamber $\blacktriangleleft F$ in X, $\phi(\blacktriangleleft F)$ lies within a uniformly bounded Hausdorff distance of a chamber in X_0.

In order to prove this, one can begin with the observation (that is proved below in 15.2.2) that $hd(\phi(F_0 \cap_p F), \bar{\phi}(F_0) \cap_{p'} \bar{\phi}(F)) < \infty$ for any $p \in F$ and $p' \in \bar{\phi}(F)$. The next step therefore is to characterize chambers among sets of type $F_0 \cap_p F$ and to prove that the map ϕ up to suitable equivalence, sends chambers to chambers. These considerations prompt the following definitions.

By a *closed chamber* or *closed chamber wall* we mean the topological closure of a chamber or chamber wall.

Let X be a metric space, and let $A \subset X$ and $B \subset X$.

DEFINITION. $A < B$ if and only if $A \subset T_v(B)$ for some finite v. $A \sim B$ if $A < B$ and $B < A$.

Thus $A \sim B$ *if and only if* $hd(A, B) < \infty$.

DEFINITION. Let X be a symmetric Riemannian space of the form G/K with G a semi-simple analytic linear group and K a maximal compact subgroup, and let $r = \text{rank } X$. A *splice* in X is a subset of the form $F_0 \cap_p F$ where F_0 and F are r-flats in X and $p \in F$.

We have seen in Lemma 7.5 that a splice is a convex union of chambers or chamber walls. A splice C is called *irreducible* if given splices $C_1, C_2, \ldots, C_n$ with $C \sim C_1 \cup C_2 \cup \ldots C_n$, then $C \sim C_i$ for some i.

LEMMA 15.1. *A splice is irreducible if and only if it is a closed chamber or chamber wall.*

Proof. The "if" part of our assertion follows at once from the fact that a closed simplex σ is not the union of closed proper subsimplices (whose vertices are vertices of σ), together with the fact that a chamber wall $\blacktriangleleft S$ lying in $T_v(\blacktriangleleft F)$ is within finite Hausdorff distance of a wall of $\blacktriangleleft F$, by Corollary 7.4.1 and Lemma 7.2 (ii).

It remains to prove the "only if" assertion. This amounts to the assertion that each closed chamber or chamber wall is a splice.

Let $\blacktriangleleft S$ be a chamber or chamber wall. Let F be an r-flat containing $\blacktriangleleft S$, set $A = pol\, F$, and let $\Phi(\blacktriangleleft S)$ denote the set of all roots α on A which are positive on $pol \blacktriangleleft S$, that is, $\alpha(pol \blacktriangleleft S) > 1$. Then $U(\blacktriangleleft S) = \Pi_\alpha G_\alpha (\alpha \in \Phi(\blacktriangleleft S))$ (cf. 2.4) and $G_F = \text{Norm}\, A$ (cf. Lemma 5.1). Let g be an element in $U(\blacktriangleleft S)$ having a non-trivial component in each G_α, $\alpha \in \Phi(\blacktriangleleft S)$. Then $g \in P(\blacktriangleleft S)$ but $g \notin P(\blacktriangleleft S_0)G_F$ (cf. 2.8) if $P(\blacktriangleleft S_0)$ does not contain $P(\blacktriangleleft S)$. As a result, $gF \cap_p F = \blacktriangleleft S \cup \blacktriangleleft S_1 \cup \ldots \cup \blacktriangleleft S_k$, p is the origin of $\blacktriangleleft S$, by Lemma 7.5 (i), and $\blacktriangleleft S_1, \ldots \blacktriangleleft S_k$ are the walls $\blacktriangleleft S_i$ such that $P(\blacktriangleleft S_i) \supset P(\blacktriangleleft S)$; that is, $\blacktriangleleft S_i$ is a face of $\blacktriangleleft S$ by 2.4. Thus the topological closure of $\blacktriangleleft S$ is a splice.

Inasmuch as any splice is a union of chambers and chamber walls by Lemma 7.5 (i), it follows at once that the only irreducible splices are the closures of chambers and of chamber walls.

THEOREM 15.2. *Let G be a semi-simple analytic group, let X denote its associated symmetric Riemannian space and let Γ be a discrete subgroup of G such that G/Γ has finite measure. Let G', X', Γ' be another such triple, and let $\theta : \Gamma \to \Gamma'$ be an isomorphism. Assume*

$\phi : X \to X'$ *is a* (k, b)*-pseudo-isometry satisfying* $\phi(\gamma x) = \theta(\gamma)\phi(x)$ for all $\gamma \in \Gamma$, $x \in X$. *Assume also the existence of a pseudo-isometry of* X' *to* X *equivariant with respect to* θ^{-1}. *Then* ϕ *induces a homeomorphism* $\phi_0 : X_0 \to X'_0$ *of the maximal boundary* X_0 *onto* X'_0, *and* $\phi_0(\gamma x) = \theta(\gamma)\phi_0(x)$ *for all* $\gamma \in \Gamma$, $x \in X_0$.

Proof. By Lemma 14.1, there is a positive constant v and a map $\overline{\phi}$ sending r-flats of X to r-flats of X' such that for any r-flat F in X

$$\text{hd}(\overline{\phi}(F), \phi(F)) < v \ . \tag{15.2.1}$$

For any r-flats F_1 and F in X and for any $p \in F$, we have by Theorem 7.8

$$F_0 \cap_p F \subset T_s(F_0) \cap F \subset T_t(F_0 \cap_p F)$$

for any $s > d(p, F_0)$ and for some finite t depending on F, F_0, p, and s. Thus $F_0 \cap_p F \sim T_s(F_0) \cap F$ if $T_s(F_0) \cap F$ is not empty.

Inasmuch as $T_r(F_0) \cap T_s(F) \subset T_s(T_{r+s}(F_0) \cap F)$, it follows easily that for any flats F_0 and F and for any r, s, t, u

$$T_r(F_0) \cap T_s(F) \sim T_t(F_0) \cap T_u(F)$$

provided each of these intersections is not empty.

Inasmuch as ϕ is a (k, b)-pseudo-isometry for some (k, b), we have

$$\phi^{-1}(\phi(A)) \subset T_b(A)$$

for any $A \subset X$. Therefore

$$\begin{aligned} \phi(F_0 \cap_p F) &\sim \phi(T_s(F_0) \cap F) \sim \phi(T_{b+s}(F_0) \cap T_{b+r}(F)) \\ &\sim \phi(\phi^{-1}\phi(T_s(F_0)) \cap \phi^{-1}\phi(T_r(F))) \\ &= \phi(T_s(F_0)) \cap \phi T_r(F) \ . \end{aligned}$$

Clearly

$$T_{(s/k)-v}(\overline{\phi}(F)) \subset \phi(T_s(F) \subset T_{ks+v}(\overline{\phi}(F)) \ .$$

Thus for s and r sufficiently large so that $T_{(s/k)-v}(\overline{\phi}(F_0)) \cap T_{(r/k)-v}(\overline{\phi}(F))$ is not empty, we find that

$$\phi(T_s(F_0)) \cap \phi(T_r(F)) \sim T_t(\overline{\phi}(F_0)) \cap T_u(\overline{\phi}(F))$$

for any t, u such that the last term is not empty. From the above considerations it is clear that for any flats F'_0 and F',

$$T_t(F'_0) \cap T_u(F') \sim F'_0 \cap_{p'} F' \tag{15.2.2}$$

for any $p' \in F'$, provided the left side is non-empty. Consequently, we get

$$\phi(F_0 \cap_p F) \sim \overline{\phi}(F_0) \cap_{p'} \overline{\phi}(F) \tag{15.2.3}$$

for any $p' \in \overline{\phi}(F)$ and $p \in F$.

This relationship admits the following interpretation. We define the *equivalence class* $[C]$ of a splice C as the set of all splices S such that $S \sim C$, that is $hd(S,C) < \infty$. Let $\mathcal{C}(X)$ denote the set of equivalence classes of splices in X. Then define $\phi^*: \mathcal{C}(X) \to \mathcal{C}(X')$ by

$$\phi^*[F_0 \cap_p F] = [\overline{\phi}(F_0) \cap_{p'} \overline{\phi}(F)] . \tag{15.2.4}$$

Since ϕ is a pseudo-isometry, $\phi(A) \sim \phi(B)$ if and only if $A \sim B$. Therefore ϕ^* is a single-valued map. Moreover, since $\overline{\phi}: \mathcal{F} \to \mathcal{F}'$ is bijective by Lemma 14.1, it follows at once that ϕ^* is bijective.

It is clear from definitions that a splice C is irreducible if and only if any splice in $[C]$ is irreducible. Moreover a splice C is irreducible if and only if any splice in $\phi^*[C]$ is irreducible. Otherwise we would find that C is an irreducible splice, and $C \sim \phi^{-1}(C'_1) \cup \ldots \cup \phi^{-1}(C'_k)$ where $C'_1, \ldots, C'_k$ are splices in X with $C'_1 \cup \ldots \cup C'_k$ not equivalent to any C'_i. But from (15.2.3) it follows that for any splice C' in X' there is a splice C in X such that $C \sim \phi^{-1}(C')$. Therefore we would get the contradiction $C \sim C_1 \cup \ldots \cup C_k$ with C not equivalent to any C_i.

By Lemma 15.1, the irreducible splices are precisely the closed chambers and chamber walls. Therefore ϕ^* sets up a correspondence between the subsets of $\mathcal{C}(X)$ and $\mathcal{C}(X')$ determined by closed chambers. Since a closed chamber may be characterized as a maximal irreducible splice, we obtain in this way a map of the set of equivalence classes of chambers in X to the set of equivalence classes of chambers of X′. That is,

(15.2.5) ϕ^* maps X_0 to X'_0 bijectively.

Let ϕ_0 denote the restriction of ϕ^* to X_0.

We must prove that ϕ_0 is continuous.

We have seen in Section 4 that G operates transitively on X_0 and $X_0 = G/G_{[\blacktriangleleft F]}$. By definition, the topology of X_0 is the quotient space topology. By Lemma 4.1 (cf. Lemma 7.1), $G_{[\blacktriangleleft F]} = P(\blacktriangleleft F)$. Consequently X_0 is compact by 2.7.

On the other hand, we can define a topology on X_0 via "uniform convergence on compact sets":

For each chamber $\blacktriangleleft F$ of origin p, define a neighborhood $U(\blacktriangleleft F, s, e)$ of $[\blacktriangleleft F]$ as the set of all $x \in X_0$ such that x has a representative $\blacktriangleleft F_0$ satisfying

$$hd(\blacktriangleleft F \cap B_s, \blacktriangleleft F_0 \cap B_s) < e ,$$

B_s denoting the ball of radius s in X and center p. The identity map of x_0 is a continuous map from the quotient topology to this second topology. Since X_0 is compact, the two topologies are equivalent. We prove next

(15.2.6) *There is a constant* c *satisfying: for each chamber* $\blacktriangleleft F$ *in* X *we can find a chamber* $\blacktriangleleft F'$ *in* X′ *such that*

$$hd(\phi(\blacktriangleleft F), \blacktriangleleft F') < c .$$

Proof. Let $\blacktriangleleft F$ be a chamber in the flat subspace F of X, let p denote the origin of $\blacktriangleleft F$, and set $\mathcal{F}(\blacktriangleleft F) = \{F_0 \in \mathcal{F};\ F_0 \cap_p F = \blacktriangleleft F\}$.

Then for any r-flat F_0 in $\mathcal{F}(\blacktriangleleft F)$, $F_0 \cap_p F$ is a chamber. Hence by (15.2.4) and (15.2.5) $\overline{\phi}(F_0) \cap_{p'} \overline{\phi}(F)$ is a chamber for any point $p' \epsilon \overline{\phi}(F)$.

We denote by $T_s^*(\ )$ the closed tube of radius s around ( ), and we note that

$$\blacktriangleleft F \subset T^*_{d(p,F_0)}(F_0) \cap F$$

by Lemma 7.5 (ii). Therefore

$$\begin{aligned}\phi(\blacktriangleleft F) &\subset \phi(\textstyle\bigcap_{F_0} T^*_{d(p,F_0)}(F_0), \quad \{F_0 \epsilon \mathcal{F}(\blacktriangleleft F)\} \\ &\subset \textstyle\bigcap_{F_0} \phi(T^*_{d(p,F_0)}(F_0), \quad \{F_0 \epsilon \mathcal{F}(\blacktriangleleft F)\} \\ &\subset \textstyle\bigcap_{F_0} T^*_{k\,d(p,F_0)}\phi(F_0), \quad \{F_0 \epsilon \mathcal{F}(\blacktriangleleft F)\}\end{aligned}$$

since ϕ is a (k,b) pseudo-isometry. Applying (15.2.1) we get

$$\phi(\blacktriangleleft F) \subset \textstyle\bigcap_{F_0} T^*_{k\,d(p,F_0)+v}\,\overline{\phi}(F_0), \{F_0 \epsilon \mathcal{F}(\blacktriangleleft F)\}\ , \tag{15.2.7}$$

For each r-flat $F_0 \epsilon \mathcal{F}$, we set $F'_0 = \overline{\phi}(F_0)$. Since ϕ is a (k,b) pseudo-isometry, we have for each r-flat F_0 in X,

$$k^{-1}\,d(p,F_0) \le d(\phi(p),\phi(F_0)) \le k\,d(p,F_0), \text{ if } d(p,F_0) \ge b\ .$$

Thus by (15.2.1)

$$k^{-1}\,d(p,F_0) - v \le d(\phi(p),F'_0) \le k\,d(p,F_0) + v\ .$$

Restricting the choice of F_0 so that $d(p,F_0) > 2kv$ as well as $d(p,F_0) \ge b$, we get

$$(2k)^{-1}\,d(p,F_0) \le d(\phi(p),F'_0)) \le 2k\,d(p,F_0), d(p,F_0) > 2kv\ . \tag{15.2.8}$$

For such r-flats $F_0 \epsilon \mathcal{F}(\blacktriangleleft F)$, the ratio

$$\frac{k\,d(p,F_0) + v}{d(\phi(p),F'_0))} \le \frac{(k + (2k)^{-1})\,d(p,F_0)}{(2k)^{-1}\,d(p,F_0)} \le 2k^2 + 1$$

if $d(p, F_0) > \sup\{2kv, b\}$. Set $e = 2k^2 + 1$ and $p'' = \phi(p)$. Then from (15.2.7) and (15.2.1) we get

$$\phi(\blacktriangleleft F) \subset T_v^*(F') \cap \bigcap_{F'_0} T^*_{e\, d(p'', F'_0)}(F'_0) \tag{15.2.9}$$

where F'_0 ranges over all elements of $\bar{\phi}(\mathcal{F}(\blacktriangleleft F))$ such that $d(p, F_0) > \sup(2kv, b)$. A fortiori (15.2.9) holds if F'_0 ranges over the elements in $\bar{\phi}(\mathcal{F}(\blacktriangleleft F))$ with $d(p'', F'_0) > 2k\, d(p, F_0)$, by (15.2.8); that is, $d(p'', F'_0) > 2k \sup\{2kv, b\}$. Let p' be a point in F' with

$$d(p', \phi(p)) \leq v \; . \tag{15.2.10}$$

Then for any $F'_0, F'_1 \in \bar{\phi}(\mathcal{F}(\blacktriangleleft F))$,

$$F'_0 \cap_{p'} F' = F'_1 \cap_{p'} F'$$

by (15.2.4); set $\blacktriangleleft F' = F'_0 \cap_{p'} F'$, F_0 in $\mathcal{F}(\blacktriangleleft F)$. Then $\bar{\phi}\, \mathcal{F}(\blacktriangleleft F) = \mathcal{F}(\blacktriangleleft F')$.

By (15.2.9),

$$\phi(\blacktriangleleft F) \subset T_v^*(F') \cap \bigcap_{F'_0} T^*_{2e\, d(p', F'_0)}(F'_0) \tag{15.2.11}$$

where F'_0 ranges over all elements of $\mathcal{F}(\blacktriangleleft F')$ such that $d(p', F'_0) > \sup\{4k^2v + v, 2kb + v\}$. On the other hand, by Theorem 7.8, there is a constant c_1 such that for any r-flat F', point $p' \in F'$, and chamber $\blacktriangleleft F'$ in F' of origin p', the right side of (15.2.11) is a subset of $T_{c_1}(\blacktriangleleft F')$; namely any constant c_1 which is valid for a single $\blacktriangleleft F'$ is valid for *all* $\blacktriangleleft F'$ since the right side of (15.2.11) is given by a covariant functor of $\blacktriangleleft F'$. Consequently for *all* chambers $\blacktriangleleft F$ in X, there is a chamber $\blacktriangleleft F'$ in X′ such that

$$\phi(\blacktriangleleft F) \subset T_{c_1}(\blacktriangleleft F') \; .$$

The opposite inclusion is proved by a similar argument. Namely, for any $s > 0$ and for any $F_0 \in \mathcal{F}$, $T_s(F'_0) \subset T_{s+v}(\phi(F_0)) \subset \phi(T_{k(s+v)}(F_0))$. Hence $\bigcap_{F'_0} T^*_{d(p', F'_0)}(F'_0) \subset \bigcap_{F'_0} \phi(T^*_{k(d(p', F'_0)+v)}(F_0)) \subset \phi(\bigcap_{F'_0} T^*_{k(d(p', F'_0)+v)+b}(F_0))$ where F'_0 can vary over any subset of $\mathcal{F}'$, and $p' \in F'$. Hence

$$\blacktriangleleft F' \subset \phi(\bigcap_{F_0} T^*_{k(d(p',F'_0)+v)+b}(F_0) \cap T_{kv+b}(F))$$

where F_0 varies over all elements $F_0 \in \mathcal{F}(\blacktriangleleft F)$ with $d(p, F_0) > 2kv + b$. For this subset S of $\mathcal{F}(\blacktriangleleft F)$ we get

$$\frac{k(d(p', F'_0)+v)+b}{d(p, F_0)} \leq \frac{k(k\,d(p, F_0)+v+v)+b}{d(p, F_0)}$$

$$\leq k^2 + 1 \; .$$

Select a constant c_2 so that

$$\bigcap_{F_0 \in S} T^*_{(k^2+1)d(p,F_0)}(F_0) \cap T_{kv+b}(F) \subset T_{c_2}(\blacktriangleleft F) \; . \tag{15.2.12}$$

Then $\blacktriangleleft F' \subset \phi(T_{c_2}(\blacktriangleleft F)) \subset T_{kc_2}(\phi(\blacktriangleleft F))$ is valid for all chambers $\blacktriangleleft F$ since the left side of (15.2.12) is given by a covariant functor of $\blacktriangleleft F$.

Set $c = \sup\{c_1, kc_2\}$. Then for all chambers $\blacktriangleleft F$ in X,

$$hd(\phi(\blacktriangleleft F), \blacktriangleleft F') < c \tag{15.2.13}$$

where $\blacktriangleleft F' = F'_0 \cap_{p'} F'$ with $d(p', \phi(p)) \leq v$.

Finally, we can verify that the map ϕ_0 is continuous. Let $x_n \in X_0$ and suppose $x_\infty = \lim_{n \to \infty} x_n$. Let $p \in X$ and for each n, let $\blacktriangleleft F_n$ be a chamber of origin p in the class $x_n (n = 1, 2, \ldots, \infty)$. Let B_r and B'_r denote the balls of radius r with center p and $\phi(p)$ in X and X′ respectively.

For each n, select $\blacktriangleleft F'_n$ as in (15.2.13) with origin p'_n in the closure of B'_v. Next select $\blacktriangleleft F''_n$ to be the chamber with origin $\phi(p)$ that is equivalent to $\blacktriangleleft F'_n$ $(n = 1, 2, \ldots, \infty)$. Since the topology of X_0 is given by "uniform convergence on compact sets," it remains to prove that for each $s > 0$

$$hd(\blacktriangleleft F''_n \cap B'_s, \blacktriangleleft F''_\infty \cap B'_s) \to 0$$

as $n \to \infty$.

By Lemma 4.2,

$$\text{(15.2.14)} \qquad hd(\blacktriangleleft F'_n, \blacktriangleleft F''_n) \leq d(p'_n, \phi(p)) \leq v$$

for $n = 1, 2, \ldots, \infty$. By hypothesis, for each $s > 0$

$$hd(\blacktriangleleft F_n \cap B_s, \blacktriangleleft F_\infty \cap B_s) \to 0$$

as $n \to \infty$. Since ϕ is a pseudo-isometry

$$hd(\phi(\blacktriangleleft F_n) \cap B'_s, \phi(\blacktriangleleft F_\infty) \cap B'_s) \to 0$$

as $n \to \infty$. By (15.2.13), therefore for each $s > 0$

$$hd(\blacktriangleleft F'_n \cap B'_s, \blacktriangleleft F'_\infty \cap B'_s) < 2c$$

for all large n. Hence by (15.2.14) for each $s > 0$

$$\text{(15.2.15)} \qquad hd(\blacktriangleleft F''_n \cap B'_s, \blacktriangleleft F''_\infty \cap B'_s) < 2c + 2v$$

for all large n. Inasmuch as $\blacktriangleleft F''_n$ and $\blacktriangleleft F''_\alpha$ have a common origin, it follows readily from Lemma 3.2 that just as in Euclidean space (15.2.15) implies for each $s > 0$

$$hd(\blacktriangleleft F''_n \cap B'_s, \blacktriangleleft F''_\infty \cap B'_s) \to 0$$

as $n \to \infty$. The proof that ϕ_0 is continuous is now complete.

The injectivity of ϕ_0 follows from the injectivity of the map ϕ^* of (15.2.4).

Since ϕ_0 is an injective continuous map of the compact manifold X_0 and has a dense image by Lemma 8.5, it is surjective (one can also see this directly) and a homeomorphism.

Finally, one can verify (by the same argument used in proving (14.1.5)) that

$$\phi_0(\gamma x) = \theta(\gamma)\phi_0(x)$$

for all $\gamma \in \Gamma$, $x \in X_0$. The proof of Theorem 15.2 is now complete.

Equation (15.2.4) above has a remarkable interpretation in terms of parabolic subgroups. For each chamber or chamber wall $\blacktriangleleft S$, the set of all splices equivalent to $\blacktriangleleft S$ consists of irreducible splices and therefore consists of closed chambers or chamber walls only by Lemma 15.1. On the other hand, we have seen above (cf. (15.2.4) and its sequel) that the map $\phi^* : \mathcal{C}(X) \to \mathcal{C}(X')$ takes equivalence classes of irreducible splices to equivalence classes of irreducible splices. Let $\blacktriangleleft S$ be a chamber or chamber wall and let $[\blacktriangleleft S]$ be the set of all chambers or chamber walls $\blacktriangleleft S_1$ with $\mathrm{hd}(\blacktriangleleft S, \blacktriangleleft S_1) < \infty$.

Let $\mathcal{S}(X)$ denote the set of all equivalence classes $[\blacktriangleleft S]$ as $\blacktriangleleft S$ ranges over all chambers and chamber walls in X. Let $\mathcal{T}(G)$ denote the set of all parabolic subgroups of G. The map $P : \blacktriangleleft S \to P(\blacktriangleleft S)$ defined in Section 7 yields a single valued map $[\blacktriangleleft S] \to P(\blacktriangleleft S)$. For by Lemma 7.2 (ii), $\blacktriangleleft S_1 \in [\blacktriangleleft S]$ if and only if $\blacktriangleleft S_1 = g\blacktriangleleft S$ with $g \in P(\blacktriangleleft S)$ so that applying the definitions of Section 7 and 2.4, we get

$$\begin{aligned} P(\blacktriangleleft S_1) &= P(g\blacktriangleleft S) = P(pol\ g\blacktriangleleft S) = P(g\ pol\ S\ g^{-1}) = g\ P(pol\blacktriangleleft S)g^{-1} \\ &= g\ P(\blacktriangleleft S)g^{-1} = P(\blacktriangleleft S)\ . \end{aligned}$$

Inasmuch as every parabolic subgroup has the form $P(\blacktriangleleft S)$ for some chamber or chamber wall (cf. 2.4) we see that P yields a bijective map of $\mathcal{S}(X)$ onto $\mathcal{T}(G)$; we call this map P_X.

Let $\phi_X : \mathcal{S}(X) \to \mathcal{S}(X')$ denote the bijective map induced by ϕ^*. Set

$$\phi_G = P_{X'}\, \phi_X\, P_X^{-1}\ .$$

Then $\phi_G : \mathcal{T}(G) \to \mathcal{T}(G')$ is a bijective map. By the same argument used in proving (14.1.5), one sees that ϕ^* is θ-equivariant. Therefore ϕ_G is θ-equivariant. The map P_X is a G-map since $P(g\blacktriangleleft S) = g\ P(\blacktriangleleft S)g^{-1}$; similarly $P_{X'}$ is a G'-map. It follows at once that the map ϕ_G is θ-equivariant; that is,

$$\phi_G(\gamma \cdot P) = \theta(\gamma) \cdot \phi_G(P) \tag{15.3.0}$$

for all $\gamma \in \Gamma$, $P \in \mathcal{T}(G)$, where $\gamma \cdot P = \gamma\, P\, \gamma^{-1}$.

LEMMA 15.3. *We continue the hypotheses of Theorem 15.2. For any parabolic subgroups* P_1 *and* P_2 *in* $\mathcal{T}(G)$,

(i) $P_1 \cap P_2 \in \mathcal{T}(G)$ *if and only if* $\phi_G(P_1) \cap \phi_G(P_2) \in \mathcal{T}(G')$.

(ii) *If* $P_1 \cap P_2 \in \mathcal{T}(G)$, *then* $\phi_G(P_1 \cap P_2) = \phi_G(P_1) \cap \phi_G(P_2)$.

(iii) ϕ_G *is* θ*-equivariant.*

Proof. Let $◂S$ and $◂S_0$ be chambers or chamber walls in X. By Corollary 7.4.1 and (2.4)

(15.3.1) $$P(◂S) \subset P(◂S_0) \text{ if and only if } ◂S_0 < ◂S .$$

We define a *partial ordering* on the set $\mathcal{S}(X)$ by:

(15.3.2) $[◂S_0] \leq [◂S]$ *if and only if* $◂S_0 < ◂S$ *for some* $◂S_0 \in [◂S_0]$ *and* $◂S \in [◂S]$.

Inasmuch as ϕ is a pseudo-isometry, we see that $\phi(◂S_0) < \phi(◂S)$ if and only if $◂S_0 < ◂S$. Moreover, by (15.2.4) and the definition of ϕ_X, we have $hd(\phi(◂S), ◂S') < \infty$ for any chamber wall in the equivalence class $\phi_X([◂S])$. It follows at once that the map ϕ_X preserves the partial order (15.3.2). In view of (15.3.1), ϕ_G preserves inclusion relations between parabolic subgroups. Similarly ϕ_G^{-1} preserves inclusions. From this and (15.3.0) Lemma 15.3 follows.

The map $\phi_G : \mathcal{T}(G) \to \mathcal{T}(G')$ is of course intimately related to the boundary map ϕ_0, and the relation between the maps can be stated as follows. For any parabolic subgroup $P \in \mathcal{T}(G)$, we define its support in X_0 via:

(15.4.1) $$|P| = \{x \in X_0;\ G_x \subset P\} .$$

Thus

(15.4.2) $$|\phi_G(P)| = \{x \in X'_0;\ G'_x \subset \phi_G(P)\} .$$

From (15.2.4) applied to equivalence classes of closed chambers we see that for all $x \in X_0$,

$$\phi_G(G_x) = G'_{\phi_0(x)} \cdot$$

We recall from Section 4 that G_x is a minimal parabolic subgroup for each $x \in X_0$ and that $x = [\ F]$ for some chamber F in X. Inasmuch as ϕ_G preserves inclusion relations, (15.4.2) may be restated as:

$$(15.4.3) \qquad |\phi_G(P)| = \phi_0(|P|) \quad \text{for all} \quad P \in \mathcal{T}(G) .$$

That is, ϕ_0 sends supports of parabolic subgroups to supports of parabolic subgroups. It is clear from the definition of supports that $|P_1 \cap P_2| = |P_1| \cap |P_2|$ if P_1, P_2, and $P_1 \cap P_2$ are parabolic subgroups. Thus Lemma 15.3 can be restated in the equivalent form:

$$(15.4.4) \qquad |\phi_0(P_1)| \cap |\phi_0(P_2)| = \phi_0(|P_1 \cap P_2|)$$

for any parabolic subgroups P_1 and P_2, where we define $|P|$ for *any* subgroup P by formula (15.4.1).

LEMMA 15.4. *Under the hypotheses of Theorem* 15.2,

$$\phi'_0 = \phi_0^{-1} .$$

Proof. By Lemma 8.5, every Γ orbit on X_0 is dense in X_0. It follows that both ϕ_0 and ϕ'_0 are uniquely determined by their effect on a single point.

By Proposition 12.6 and (11.3.ii)$'$, there is an element $\gamma \in \Gamma$ such that both γ and $\theta(\gamma)$ are polar regular. By Lemma 5.2 (ii), both γ and $\theta(\gamma)$ stabilize unique r-flats. This implies that γ stabilizes exactly w points on X_0 where w is the number of chambers of same origin in an r-flat of X – or equivalently, the order of the little Weyl group of a maximal polar subgroup of G. But even more is true: among these w fixed points, there is only one in whose neighborhood γ is contractive, i.e., sends an arbitrarily small neighborhoods into proper subset of itself. Namely, the point stabilized by $P(\blacktriangleleft A)$ where $\blacktriangleleft A$ is the unique chamber in G containing *pol* γ. Let x_γ denote the unique fixed point of γ on X_0 in whose neighborhood γ is contractive. Similarly let $x'_{\theta(\gamma)}$ be the unique such point for $\theta(\gamma)$ on X'_0. Then clearly

$$\phi_0(x_\gamma) = x'_{\theta(\gamma)} \quad \text{and} \quad \phi'_0(x'_{\theta(\gamma)}) = x_\gamma \ .$$

Therefore $\phi'_0 = \phi_0^{-1}$ on $\overline{\Gamma' x'_{\theta(\gamma)}} = x'$.

§16. Tits Geometries

In his famous Erlangen program, Felix Klein proposed to study the classical geometries by means of their groups of automorphisms. For example, to each point or line x of projective n-space, one can assign its stabilizer subgroup G_x in the projective linear group G. In order to study incidence relations among k-planes $(k = 0, 1, 2, \ldots, n)$ one introduces maximal subsets of mutually incident subsets or *flags* $L_0 \subset L_1 \subset \ldots \subset L_n$ where each L_k is a k-plane. The k-plane L and k'-plane L' are incident if and only if they can be included in a flag F; or equivalently if and only if $G_L \cap G_{L'} \supset G_F$. Now comes a key observation. The stabilizer of a flag is a parabolic subgroup and a fortiori, the stabilizer of a k-plane is a parabolic subgroup. Thus the k-plane L and the k'-plane L' are incident if and only if the intersection of their stabilizers is a parabolic subgroup.

With this observation as a starting point, Jacques Tits has succeeded in carrying out the Erlangen program in reverse. Namely, to each algebraic group G, or more generally to each group satisfying his "B-N pair" axioms, he associates a geometry $\mathfrak{J}(G)$ consisting of the parabolic subgroup of G. Two elements P and P' in $\mathfrak{J}(G)$ are called *incident* if and only if $P \cap P' \in \mathfrak{J}(G)$. Tits then studies the group G by means of the geometry $\mathfrak{J}(G)$.

For our purpose we require the generalization of the Fundamental Theorem of Projective Geometry for Tits geometries. As it happens, the proof of the precise theorem we require can be extracted from Tits' account (cf. [20], Theorem 5.18.4) but the statement of his Theorem 5.18.4 has a hypothesis that is too restrictive for the application we have in mind: Tits assumes in his Theorem 5.18.4 that G is "absolutely simple of k-rank ≥ 2'' whereas in our application G is semi-simple with all factors of k-rank ≥ 2.

In order to convey the flavor of Tits' theory, we sketch the proof of the theorem we need (Theorem 16.1 below), prefacing it with a few explanatory remarks about the terms that occur in the theorem.

Let G be an algebraic linear group defined over a field k and let $\mathcal{J}(G,k)$ denote the set of all k-parabolic subgroups of G; that is, those defined over k. Let S be a maximal k-split torus in G and let Σ_S be the set of all k-parabolic subgroups containing S. Let $\mathfrak{A}$ denote all subsets of $\mathcal{J}(G,k)$ of the form Σ_S for some k-split torus S. Let $\Delta(G,k)$ denote the pair $(\mathcal{J}(G,k),\mathfrak{A})$. Then $\Delta(G,k)$ is called the *building* attached to (G,k); the elements of $\mathfrak{A}$ are called *apartments*. The minimal k-parabolic subgroups are called *chambers* in $\mathcal{J}(G,k)$. If a is a root on S and $U_{(a)}$ denotes the unipotent "root subgroup" of a, then the subset of all parabolic subgroups in Σ_S which contain $U_{(a)}$ is called a *root* of the apartment Σ_S.

We have defined the notion of building $\Delta(G,k)$, apartment, root, and chamber above in terms of the group structure of (G,k). However, Tits made the fundamental discovery that all these notions can be defined solely in terms of the partial ordering of inclusion in $\mathcal{J}(G,k)$. Actually, it is more convenient to introduce as partial ordering on $\mathcal{J}(G,k)$ the relation

$$C_1 \leq C_2 \quad \text{if and only if} \quad C_1 \supseteq C_2 \ .$$

In this situation we say "C_1 is a face of C_2." This partial ordering being understood, a *chamber* in $\mathcal{J}(G,k)$ can be defined as a *maximal* element. Two distinct chambers C_1 and C_2 are called *adjacent* if and only if there is an element C with $C < C_1$, $C < C_2$, and $C \leq D < C_i$ implies $C = D$ both for $i = 1$ and $i = 2$. The notions above can also be defined in terms of the partial ordering, but their definitions depend on more intricate features of the partial ordering in a Tits building.

In view of our results in Section 15, the question we wish to answer is: Suppose the Tits building associated to two groups are isomorphic. Does this isomorphism induce an isomorphism of the groups?

Before pursuing this question further, we must make a slight modification in our definition of $\Delta(G,k)$ so as to accommodate semi-simple analytic linear groups.

Let G be a semi-simple linear analytic group in $GL(n,\mathbf{R})$. Let G^* denote the Zariski-closure of G in $GL(n,\mathbf{C})$. Then $G = (G^*\mathbf{R})^0$, the Euclidean-topologically connected component of the identity in $G_{\mathbf{R}} = G^* \cap GL(n,\mathbf{R})$, and $G_{\mathbf{R}}/G^0$ is finite. Moreover, the $\mathbf{R}$-parabolic subgroups of G^* intersect G in precisely the parabolic subgroups as defined in 2.4, and vice-versa. Thus the set $\mathcal{T}(G^*,\mathbf{R})$ coincides with the set $\mathcal{T}(G)$ introduced in Section 15; i.e.,

$$\mathcal{T}(G) = \mathcal{T}(G^*,\mathbf{R}) \ . \tag{16.1.1}$$

The next theorem will imply that if G and G' are semi-simple analytic groups having no center and no factors of $\mathbf{R}$-rank ≤ 1, then an isomorphism of $\mathcal{T}(G)$ to $\mathcal{T}(G')$ induces an isomorphism of G to G'.

Recall that the k-rank of an algebraic group defined over k is the dimension of a maximal k-split torus. If G is a semi-simple analytic linear group in $GL(n,\mathbf{R})$, a maximal polar subgroup of G is one of the form $(S_{\mathbf{R}})^0$ where S is a maximal $\mathbf{R}$-split torus of the Zariski-closure G^*.

THEOREM 16.1 (Tits, cf. [20]). *Let G be a connected semi-simple linear algebraic group defined over a field k. Assume that G has no center and no factor of k-rank ≤ 1. Let G^+ denote the subgroup of G generated by all U_k, where U is the unipotent radical of a k-parabolic subgroup and U_k denotes its k-rational points. Then G^+ is canonically determined by the Tits building $\Delta(G,k)$.*

We merely sketch Tits' proof.

Set $\Delta = \Delta(G, k)$. The group G_k operating by inner automorphisms permutes the k-parabolic subgroups of G thus giving rise to a canonical homomorphism $\tau : G_k \to \mathrm{Aut}\,\Delta$. The kernel of τ is the intersection N of all the k-parabolic subgroups of G. This intersection is an algebraic k-subgroup having no k-split torus and as a result has k-rank 0. Furthermore, N is invariant under conjugations by G_k and therefore under conjugations by G since G_k is Zariski-dense in G. Thus N is a normal subgroup of G. By our hypotheses on G, $\dim N = 0$ and thus N is discrete. Since a discrete normal subgroup of a connected group is central, $N = (1)$. Thus $\tau : G_k \to \mathrm{Aut}\,\Delta$ is an injection.

It remains to define $\tau(G^+)$ in terms of the building structure. Inasmuch as G^+ is generated by the root subgroups $U_{(\alpha),k}$ where α varies over all roots on all k-split tori, it suffices to locate the collection $\{\tau(U_{(\alpha),k})\}$ in $\mathrm{Aut}\,\Delta$.

Let Σ be an apartment in Δ and let α be a root of Σ. Since G has no factors of rank 1, one can find a chamber C in α such that each maximal face of C is a face of exactly two chambers in the root α. Consider now the subgroup of elements g in $\mathrm{Aut}\,\Delta$ defined by the two conditions

(1) g keeps fixed each chamber in the root α.

(2) g keeps fixed each chamber adjacent to C.

Tits proves that this subgroup coincides with $\tau(U_{(\alpha),k})$ for some root α (in the group sense) on the maximal k-split torus corresponding to an apartment containing the (building-sense) root α.

It follows at once that $\tau(G^+)$ is definable in terms of the building structure.

COROLLARY 16.2. *Let* G *and* G′ *be semi-simple analytic linear groups. Set* $\mathcal{T}(G) = \mathcal{T}(G^*, R)$, $\mathcal{T}(G') = \mathcal{T}(G'^*, R)$, $\Delta = \Delta(G^*, R)$, $\Delta' = \Delta(G'^*, R)$. *Let* $\phi : \mathcal{T}(G) \to \mathcal{T}(G')$ *be an order-preserving bijection. Assume that* G *and* G′ *have no* R*-rank* 1 *quotient groups. Then*

$$\phi \circ \tau(G) \circ \phi^{-1} = \tau(G') . \tag{16.2.1}$$

Proof. Let N denote the kernel of the canonical homomorphism $\tau : G \to \mathrm{Aut}\, \Delta$. Clearly no generality is lost in dividing G and G′ by the kernel N and the kernel of $\tau : G' \to \mathrm{Aut}\, \Delta'$ respectively. Since N is the intersection of all the parabolic subgroups of G, it contains the center of G as well as the maximum normal $\mathbf{R}$-rank 0 subgroup of G. Thus, upon replacing G by G/N, we have that G is center free and has no $\mathbf{R}$-rank 0 factor. The hypotheses of the Corollary imply that G and G′ have no factors of $\mathbf{R}$-rank ≤ 1. Therefore we may apply Theorem 16.1 to conclude

$$\phi \circ \tau(G^+) \circ \phi^{-1} = \tau(G'^+) \ .$$

Now for any semi-simple analytic group G, we have $G = N G^+$ where N is the maximum normal compact subgroup – or equivalently, $\mathbf{R}$-rank 0 subgroup. Thus in our situation, $G = G^+$ and $G' = G'^+$. It follows at once that

$$\phi \circ \tau(G) \circ \phi^{-1} = \tau(G') \ .$$

§17. Rigidity for R-rank > 1

In Section 15, we have proved that the boundary map induces an isomorphism of the Tits geometry of G onto the Tits geometry of G′. In the case of R-rank greater than one, we can apply Corollary 16.2 to deduce our rigidity theorem. But we must make some preliminary remarks in order to match up the hypotheses of the Theorems that we wish to invoke.

LEMMA 17.1 (Selberg). *Let* Γ *be a finitely generated subgroup of* $GL(n,C)$. *Then* Γ *contains a torsion-free subgroup* Γ_1 *of finite index.*

This is proved in [16]. A more detailed discussion is given by Borel in [2b].

It should be noted that the subgroup Γ_1 can be taken normal in Γ – for one can replace it by $\bigcap_{\gamma \epsilon \Gamma} \gamma \Gamma_1 \gamma^{-1}$.

THEOREM 17.2. *Let* G *and* G′ *be semi-simple analytic groups having no center and no factors of* R*-rank* ≤ 1. *Let* Γ *and* Γ' *be discrete subgroups of* G *and* G′ *respectively such that* G/Γ *and* G'/Γ' *are compact. Let* $\theta : \Gamma \to \Gamma'$ *be an isomorphism. Then there is an analytic isomorphism*

$$\overline{\theta} : G \to G'$$

such that θ *is the restriction of* $\overline{\theta}$ *to* Γ.

Proof. We first consider the case that Γ is torsion-free.

Let K and K′ denote maximal compact subgroups in G and G′ respectively. Set $X = G/K$ and $X' = G'/K'$. By Lemma 9.2, there exists a θ-equivariant pseudo-isometry

$$\phi : X \to X'$$

and a θ^{-1}-equivariant pseudo-isometry

$$\phi' : X' \to X \ .$$

Let X_0 and X'_0 denote the maximal boundaries of X and X' respectively (cf. Section 4). By Theorem 15.2, there is a θ-equivariant homeomorphism $\phi_0 : X_0 \to X'_0$. By Lemma 15.3, there is an order-preserving bijection $\phi_G : \mathfrak{T}(G) \to \mathfrak{T}(G')$. There is a canonical map $X_0 \to \mathfrak{T}(G)$ given by $x \to G_x$, since the stabilizer of each point in X_0 is a parabolic subgroup. Identifying X_0 with its image in $\mathfrak{T}(G)$, we see by (15.4.3) that ϕ_0 is the restriction of ϕ_G to X_0.

Let τ denote the canonical map of $G \to \mathrm{Aut}\, \mathfrak{T}(G)$ and $G' \to \mathrm{Aut}\, \mathfrak{T}(G')$. By Corollary 16.2

$$\phi_G \circ \tau(G) \circ \phi_G^{-1} = \tau(G') \ .$$

For each $g \in G$, let $\tau_0(g)$ denote the restriction of $\tau(g)$ to X_0. Since X_0 and X'_0 are stable under $\tau(G)$ and $\tau(G')$ respectively, we get

$$\phi_0 \circ \tau_0(G) \circ \phi_0^{-1} = \tau_0(G') \ . \tag{17.2.1}$$

Inasmuch as the intersection of all the minimal parabolic subgroups in G (resp. G') are factors of $\mathbb{R}$-rank 0, we infer that the map $\tau_0 : G \to \mathrm{Aut}\, X_0$ (resp. $G' \to \mathrm{Aut}\, X'_0$) is injective. Thus (17.2.1) defines an isomorphism $\overline{\theta} : G \to G'$. By Theorem 15.2, the map $\phi_0 : X_0 \to X'_0$ is a homeomorphism. Inasmuch as X_0 is a homogeneous G-space and G is a countable union of compact sets, the topology of G can be recaptured from its action on X_0. It follows that the isomorphism $\overline{\theta} : G \to G'$ is analytic, by a well-known theorem of E. Cartan (cf. [8b]). By Theorem 15.2, the map ϕ_0 is θ-equivariant. That is, for all $\gamma \in \Gamma$ and $x \in X_0$

$$\phi_0(\gamma x) = \theta(\gamma)\phi_0(x) \ .$$

In other words, for all $\gamma \in \Gamma$

$$\tau_0(\theta(\gamma)) = \phi_0 \circ \tau_0(\gamma) \circ \phi_0^{-1} .$$

By definition of $\overline{\theta}$, for all $g \in G$,

$$\tau_0(\overline{\theta}(g)) = \phi_0 \circ \tau_0(g) \circ \phi_0^{-1} .$$

Since τ_0 is an isomorphism, we conclude that θ is the restriction to Γ of θ. Thus Theorem 17.2 is proved in the case that Γ is torsion-free.

In the general case, we can find a normal subgroup Γ_1 of Γ of finite index by Lemma 17.1. Applying the foregoing result to Γ_1, there is an analytic isomorphism $\overline{\theta} : G \to G'$ such that θ and $\overline{\theta}$ have the same restrictions on Γ_1. Without loss of generality, we replace G' by G and θ by $\overline{\theta}^{-1} \circ \theta$; we then can assume that $\overline{\theta}$ is the identity map of G and $\theta(\gamma) = \gamma$ for all $\gamma \in \Gamma_1$. Then for any $\gamma \in \Gamma$ and $g \in \Gamma_1$, we have

$$\gamma g \gamma^{-1} = \overline{\theta}(\gamma g \gamma^{-1}) = \theta(\gamma g \gamma^{-1}) = \theta(\gamma) g \theta(\gamma)^{-1} .$$

Therefore for all $\gamma \in \Gamma$, $\theta(\gamma)^{-1}\gamma$ centralizes Γ_1. Since G/Γ_1 has finite measure, the subgroup Γ_1 is Zariski-dense in G by Lemma 8.6. Consequently $\theta(\gamma)^{-1}\gamma$ centralizes G. By hypothesis, G has no center. It follows that $\theta(\gamma)^{-1}\gamma = 1$ for all $\gamma \in \Gamma$. Therefore $\theta = \overline{\theta}$ on Γ. The proof of our theorem is now complete.

§18. The Restriction to Simple Groups

Let G be a semi-simple analytic linear group having no compact normal subgroup other than (1). Then G has no center and is therefore the direct product of all its minimal normal proper subgroups. Moreover, all of these factors have R-rank at least one. Let Γ be a discrete subgroup of G such that G/Γ is compact. In this section we shall reduce the rigidity question for (G,Γ) to the same question for the factors of (G,Γ). Our method rests on the existence of the boundary homeomorphism ϕ_0 of the maximal boundary X_0. It would therefore apply equally well in the case G/Γ merely has finite measure, provided one could establish in this case the conclusion of Lemma 9.2 on the existence of the pseudo-isometric Γ-map ϕ of the associated symmetric space X.

18.1. Let $\mathcal{T}(G)$ denote the set of all parabolic subgroups of G. Then $\mathcal{T}(G)$ is a finite union of G-orbits since each parabolic subgroup is conjugate to $P(^{\blacktriangleleft}B)$ where $^{\blacktriangleleft}B$ is a chamber or chamber wall (cf. 2.4). We endow each of the orbits with its quotient topology induced from G. Let $\mathcal{T}^1(G)$ denote the subspace of $\mathcal{T}(G)$ consisting of proper maximal parabolic subgroups. For any $P \in \mathcal{T}^1(G)$, set

$$G(P) = \bigcap_g gPg^{-1} \qquad \{g \in G\} .$$

Then G(P) is a normal subgroup of G containing all factors of G except one, since P does. Hence G/G(P) is simple and we have $G \approx (G/G(P)) \times G(P)$ as a direct product of analytic groups.

18.2. Let X_1 and X_2 be unions of G-orbits in $\mathcal{T}(G)$. We say that X_1 and X_2 are *independent* if each $P_1 \in X_1$ is incident with each $P_2 \in X_2$;

that is $P_1 \cap P_2 \in \mathcal{T}(G)$ for each $P_1 \in X_1$ and $P_2 \in X_2$. Let $X_1^1, X_2^1, \ldots, X_n^1$ be a maximal set of mutually independent G-orbits in $\mathcal{T}^1(G)$. Let $\mathcal{T}_i(G)$ denote the set of all elements in $\mathcal{T}(G)$ which are incident with each element of $\cup_{j \neq i} X_j$; $(i = 1, 2, \ldots, n)$. Then the map

$$\rho : (P_1, P_2, \ldots, P_r) \to P_1 \cap P_2 \cap \ldots \cap P_n \tag{18.2.1}$$

gives a bijection of $\mathcal{T}_1(G) \times \ldots \times \mathcal{T}_n(G)$ onto $\mathcal{T}(G)$. One can see this by observing that any Q_i in X_i has the form $Q_i = P(\blacktriangleleft B_i)$ where $\blacktriangleleft B_i$ is a one-dimensional chamber wall of a chamber $\blacktriangleleft A_i$ in a simple factor G_i of G, that $\blacktriangleleft A_1 \times \ldots \times \blacktriangleleft A_n$ is a chamber in G for any set of chambers $\blacktriangleleft A_i$ in G_i $(i = 1, \ldots, n)$.

Now although the G-orbit X_i^1 can be replaced above by any other G-orbit in $\mathcal{T}^1(G) \cap \mathcal{T}_i(G)$, the factors $\mathcal{T}_i(G)$ appearing in

$$\mathcal{T}(G) = \mathcal{T}_1(G) \times \ldots \times \mathcal{T}_n(G) \tag{18.2.2}$$

are unique up to order. Indeed they are defined uniquely, up to order, by the condition (18.2.1).

18.3. Let X_0 denote the maximal boundary of G. As in Section 17, we identify X_0 with its canonical image in $\mathcal{T}(G)$ and we let X_i denote the orbit of all minimal elements in $\mathcal{T}_i(G)$, $i = 1, 2, \ldots, n$. Then (18.2.1) yields

$$X_0 = X_1 \times X_2 \times \ldots \times X_n \ . \tag{18.3.1}$$

As a result, we can think of each X_i in two ways: either as a subspace of X_0 or as a quotient space. It is the latter viewpoint that we now pursue.

For each $P_i \in X_i$ we have $X_i = G/P_i$, $i = 1, 2, \ldots, n$.

Let $\pi_i : X_0 \to X_i$ be given by

$$\pi_i(P_1 \cap \ldots \cap P_n) = P_i \qquad i = 1, 2, \ldots, n \ ;$$

this map is well defined since the map ρ of (18.2.1) is a bijection. Then each π_i is a G-invariant fibration of X_0, the kernel of the G-operation on the base space $\pi_i(X_0)$ is $G(P_i)$ for any $P_i \in X_i$; and the diagonal map of G yields a homomorphism

$$G \to \prod_{i=1}^{n} G/G(P_i) . \qquad (18.3.2)$$

The kernel of this map is $G(P_1) \cap \ldots G(P_n)$ which is $G(P_1 \cap \ldots \cap P_n)$. Set $P_1 \cap \ldots \cap P_n = P$. P is a minimal parabolic subgroup and therefore G(P) is a $\mathbb{R}$-rank 0 normal subgroup of G; that is, a compact normal subgroup of G. Therefore $G(P) = 1$ and the map (18.3.2) is an isomorphism.

Let $\tau_i(g)$ denote the canonical operation of the element $g \in G$ on X_i $(i = 0, 1, 2, \ldots, n)$. Then $\mathrm{Ker}\, \tau_i = G(P_i)$ for any $P_i \in X_i$, $i = 0, 1, \ldots, n$ and $\tau_0 = (\tau_1, \ldots, \tau_n)$ is the isomorphism of (18.3.2). We call the decomposition (18.3.1) the "decomposition of X_0 into simple G-factors."

18.4. Let G and G′ be semi-simple analytic groups having no compact normal subgroups other than (1). Let Γ and Γ' be discrete subgroups such that G/Γ and G'/Γ' are compact. Let $\theta : \Gamma \to \Gamma'$ be an isomorphism.

By Theorem 15.2 and Lemma 15.3, there is a θ-equivariant incidence preserving bijection $\phi_G : \mathcal{T}(G) \to \mathcal{T}(G')$ which gives a homeomorphism of X_0 onto X'_0. Set $X'_i = \phi_G(X_i)$ $(i = 1, \ldots, n)$. Then

$$X'_0 = X'_1 \times X'_2 \times \ldots \times X'_n \qquad (18.4.1)$$

since ϕ_0 is a homeomorphism. We claim: *the decomposition* (18.4.1) *is the decomposition of* X'_0 *into simple* G′*-factors*.

We can justify this claim in one of two ways. The first argument uses the fact that the decomposition (18.2.2) of the Tits geometry $\mathcal{T}(G)$ into a direct product of "simple factors" is unique, up to order. That being so, the incidence preserving bijection ϕ_G sends $\mathcal{T}_i(G)$ to a simple factor $\mathcal{T}_i(G')$ of $\mathcal{T}(G')$ and therefore

$$\phi_G(X_i) = \phi_G(X_0 \cap \mathcal{T}_i(G)) = X'_0 \cap \mathcal{T}_i(G') .$$

Thus $\phi_G(X_i)$ is G'-invariant.

Our second argument is: the above decomposition $X_0 = X_1 \times \ldots \times X_r$ was defined in terms of the incidence relations in $\mathcal{T}(G)$ and certain G-orbits. We have topologized $\mathcal{T}(G)$ as a disjoint union of G-orbits. From (15.4.3) it follows easily that $\phi_G : \mathcal{T}(G) \to \mathcal{T}(G')$ is a homeomorphism. The map ϕ_G is a Γ-space map because it is θ-equivariant. By Lemma 8.5, each Γ-orbit in $\mathcal{T}(G)$ is topologically dense in its G-orbit. Therefore, wherever we used in our description of the factors $X_1, \ldots, X_n$ the term "G-orbit," we could have equally well said "closure of a Γ-orbit." Inasmuch as ϕ_G is a Γ-space map and a homeomorphism, it follows that ϕ_0 carries the decomposition of X_0 into its simple G-factors into the corresponding decompositions for X'_0.

18.5. Consider the Lie group $\tau_i(G)$ as a group of transformations of the space X_i, $i = 1, \ldots, n$. Then the Lie group topology of $\tau_i(G)$ coincides with its topology as a transformation group (cf. [8a]) and thus $\tau_i(G)$ is closed in the group of self-homeomorphisms of the space X_i, $i = 1, \ldots, n$. Let ϕ_i denote the restriction of ϕ_G to X_i; then by the result of the preceding paragraph

$$\phi_i \circ \tau_i(\gamma) \circ \phi_i^{-1} = \tau_i(\theta(\gamma))$$

for all $\gamma \in \Gamma$, $i = 1, \ldots, n$. Consequently,

$$\phi_i \circ \tau_i(\Gamma) \circ \phi_i^{-1} = \tau_i(\Gamma') \tag{18.5.1}$$

for each Γ.

We claim: the rigidity of (G, Γ) is equivalent to the assertion that for $i = 1, 2, \ldots, n$

$$\phi_i \circ \tau_i(G) \circ \phi_i^{-1} = \tau_i(G') \ . \tag{18.5.2}$$

For $\phi_0 \circ \tau_0(\gamma) \circ \phi_0^{-1} = \tau_0(\theta(\gamma))$ and $\tau_0 = (\tau_1, \ldots, \tau_n)$ imply that for all $\gamma \in \Gamma$,

$$\theta(\gamma) = \tau_0^{-1}\left(\prod_{i=1}^{n} \phi_i \circ \tau_i(\gamma) \circ \phi_i^{-1}\right) \ . \tag{18.5.3}$$

Since the right side of (18.5.3) is well defined for all $\gamma \in G$ if (18.5.2) holds, we can conclude from (18.5.2) that θ extends to a continuous (and hence analytic) isomorphism of G onto G′.

Set $G_i = \tau_i(G)$, $\Gamma'_i = \tau_i(\Gamma)$, $G'_i = \tau_i(G')$, and $\Gamma'_i = \tau_i(\Gamma')$. We distinguish two cases

Case 1. Γ_i is closed in G_i.
Case 2. Γ_i is not closed in G_i.

In case 1, Γ_i is discrete in G_i and is therefore a discrete transformation subgroup of the space X_i. Since ϕ_i is a homeomorphism, the subgroup Γ'_i is a discrete transformation subgroup of the space X'_i and is therefore discrete in G'_i. It is clear that G_i/Γ_i and G'_i/Γ'_i are compact, since G/Γ and G'/Γ' are. Since X_i is the maximal boundary of G_i, the assertion (18.5.1) is equivalent to an affirmative solution to the rigidity question for the pair (G_i, Γ_i).

In case 2, $\overline{\Gamma_i^0}$, the connected component of the identity of the Lie group closure of Γ_i is invariant under inner automorphisms of Γ_i and hence of G_i since, by Lemma 8.5, Γ_i is Zariski-dense in G_i. Since $\overline{\Gamma_i^0} \neq (1)$ and G_i is simple, we see that $G_i = \overline{\Gamma_i}$. Inasmuch as G'_i is closed in the transformation group of the space X'_i, it follows at once that Γ'_i is a dense subgroup of the closed subgroup G'_i and therefore taking the topological closure of both sides of (18.5.1), we get
$\phi_i \circ \tau_i(G) \circ \phi_i^{-1} = \tau_i(G')$.

We summarize our discussion in the following theorem.

THEOREM 18.1. *Let* G *be a semi-simple analytic group having no compact normal subgroup other than* (1). *Let* Γ *be a discrete subgroup such that* G/Γ *is compact. Let* $G = G_1 \times \ldots \times G_n$ *the decomposition of* G *into its simple factors arranged so that* $\tau_i(\Gamma)$ *is closed for* $i = 1, \ldots, s$ *and* $\tau_i(\Gamma)$ *is not closed for* $i > s$, *where* $\tau_i : G \to G_i$ *is the projection onto* G_i. *Then the pair* (G, Γ) *is rigid if and only if each pair* $(G_i, \tau_i(\Gamma))$ *is rigid for* $i \leq s$.

DEFINITION. Let (G, Γ) be as in the theorem above. The subgroup Γ is called *irreducible* in G if the image of Γ in every proper quotient $G/G_{i_1} \times \ldots \times G_{i_m}$ is not closed.

As an immediate consequence of Theorem 18.1, we get

COROLLARY 18.2. *Let* (G, Γ) *be as in Theorem* 18.1. *If* Γ *is irreducible in* G, *and if* G *is not simple, then the pair* (G, Γ) *is rigid.*

Proof. Continuing the notation above, we have $\tau_i(\Gamma)$ is not closed for every $i = 1, 2, \ldots, n$ if Γ is irreducible in G. Now apply Theorem 18.1.

REMARK. It is not hard to deduce from Lemma 8.6 that: Let $G = E \times F$ where E is a semi-simple analytic group having no infinite compact normal analytic subgroup and F is any locally compact group, and let Γ be a discrete subgroup such that G/Γ has finite measure. If ΓE is closed, then ΓF is closed, and $\Gamma \cap E \times \Gamma \cap F$ is of finite index in Γ. Thus the hypothesis in Theorem 18.1 is equivalent to:

Γ is commensurable with $(\Gamma \cap G_1) \times \ldots \times (\Gamma \cap G_s) \times \Delta$ where Δ is a discrete subgroup of $G_{s+1} \times \ldots \times G_n$ whose projection into each factor has a dense image.

§19. Spaces of R-rank 1

The symmetric Riemannian spaces of R-rank 1 and of negative curvature are the *hyperbolic spaces* $H^n_{\mathbf{K}}$, where $\mathbf{K}$ is either

(a) $\mathbf{R}$, the real numbers

(b) $\mathbf{C}$, the complex numbers

(c) $\mathbf{H}$, the quaternions

(d) $\mathbf{O}$, the Cayley members; in this case $n = 2$.

These arise as the quotient spaces

(a) $SO_0(1,n)/SO(n)$

(b) $SU(1,n)/U(n)$

(c) $Sp(1,n)/Sp(n)$

(d) $F_4/\text{Spin } 9$.

The usual Poincaré model for real hyperbolic space generalizes to the first three cases straightforwardly.

Let $x \to \bar{x}$ denote the standard involution of $\mathbf{K}$. On $\mathbf{K}^{n+1}$ we define the form

$$(19.1) \qquad \langle x, y\rangle = \bar{x}_0 y_0 - \bar{x}_1 y_1 - \bar{x}_2 y_2 - \dots - \bar{x}_n y_n \ .$$

We regard $\mathbf{K}^{n+1}$ as a right $\mathbf{K}$-module and we denote by G the subgroup of $GL(n, \mathbf{K})$ which stabilizes the form $\langle x, y\rangle$. Introduce the non-homogeneous coordinates

$$u_i = x_i x_0^{-1} \qquad (i = 1, \dots, n)$$

and on $\mathbf{K}^n$ introduce the positive definite form

$$(u, v) = \bar{u}_1 v_1 + \dots + \bar{u}_n v_n \ .$$

We take as a model for $H^n_{\mathbf{K}}$ the domain

$$D = \{u \in \mathbf{K}^n;\ (u,u) < 1\}$$

and we take as metric

$$(19.2) \qquad ds^2 = -<x,x>^{-2} \det\begin{pmatrix} <x,x> & <dx,x> \\ <x,dx> & <dx,dx> \end{pmatrix} .$$

It should be remarked that the right side of this formula is indeed well defined on D. For we have from $x_i = u_i x_0$

$$dx_i = du_i x_0 + u_i dx_0 .$$

Moreover

$$1 - (u,u) = <xx_0^{-1}, xx_0^{-1}> = \bar{x}_0^{-1} <x,x> x_0 = <x,x> \bar{x}_0^{-1} x_0^{-1} .$$

Thus

$$ds^2 = -<x, x_0^{-1}, xx_0^{-1}>^{-1} <x,x>^{-1} \det\begin{pmatrix} <xx_0^{-1}, xx_0^{-1}> & <dx, xx_0^{-1}> \\ <xx_0^{-1}, dx> & <dx,dx> \end{pmatrix}.$$

The determinant term can be expressed as

$$(1-(u,u))(d\bar{x}_0\, dx_0 - (du\cdot x_0 + u\,dx_0, du\cdot x_0 + u\,dx_0)) -$$
$$(d\bar{x}_0 - (du\cdot x_0 + u\,dx_0, u))(dx_0 - (u, du\cdot x_0 + u\,dx_0)) .$$

The coefficient of $d\bar{x}_0\, dx_0$ in this term is

$$(1-(u,u))(1-(u,u)) - (1-(u,u)-(u,u)+(u,u)^2)$$

and is thus zero. Similarly, the other parts of this term involving dx_0 are

$$-(1-(u,u))(du\cdot x_0, u\,dx_0) + u\,dx_0, du\cdot x_0)) -$$
$$(-d\bar{x}_0(u,du\cdot x_0) + (u\,dx_0,u)(u,du\cdot x_0) - (du\cdot x_0,u)\,dx_0 + (du\cdot x_0,u)(u,u\,dx_0)$$

and this also vanishes. Thus we get

$$ds^2 = -(1-(u,u))^{-1}\langle x,x\rangle^{-1}(-(1-(u,u))(du\cdot x_0, du\cdot x_0) - (du\cdot x_0,u)(u,du\cdot x_0))$$

$$= (1-(u,u))^{-1}\langle xx_0^{-1}, xx_0^{-1}\rangle^{-1}((1-(u,u))(du,du) + (du,u)(u,du)$$

$$ds^2 = (1-(u,u))^{-1}\,(|du|^2 + (1-(u,u))^{-1}\,|(du,u)|^2)\ . \tag{19.3}$$

This formula shows that (19.1) is well-defined on D. The rays issuing from the origin are geodesics for this metric and their transforms under G give the geodesics for $H^n_{\mathbf{K}}$. One can thus calculate the distance between any two points u and v of $H^n_{\mathbf{K}}$:

$$d(u,v) = \cosh^{-1} \frac{|1-(u,v)|}{|1-(u,u))^{\frac{1}{2}}\left(1-(v,v)^{\frac{1}{2}}\right)}\ . \tag{19.4}$$

In homogeneous coordinates, this formula is

$$d(x,y) = \cosh^{-1} \frac{|\langle x,y\rangle|}{\langle x,x\rangle^{\frac{1}{2}}\ \langle y,y\rangle^{\frac{1}{2}}}\ . \tag{19.4$'$}$$

The foregoing calculations break down for the case $\mathbf{K}=\mathbf{O}$, because $\mathbf{O}$ is non-associative. For example $(xz)(zy) \neq |z|^2 xy$ for general x, y, z in $\mathbf{O}$.

There is however a description of $H^2_{\mathbf{O}}$ due to Veldkamp and Springer [19], that is based on an idea of Freudenthal [5b] which is valid for all $\mathbf{K}$. Freudenthal's idea is: Instead of representing points of n-dimensional projective space $P^n_{\mathbf{K}}$ as cyclic $\mathbf{K}$-subspaces of $\mathbf{K}^{n+1}$ with respect to some inner product. In turn such projections span a Jordan algebra of hermitian matrices with respect to Jordan multiplication

$$xy = \frac{1}{2}(x\cdot y + y\cdot x)$$

where $x\cdot y$ is ordinary matrix multiplication. The success of this idea is related to the discovery by Chevalley and Schafer that F_4 is the group of automorphisms of the Jordan algebra of 3×3 hermitian matrices with entries from $\mathbf{O}$.

To be explicit, let J denote the set of all 3×3 matrices x with coefficients in $\mathbb{O}$ that are hermitian with respect to the inner product on $\mathbb{O}^3$:

$$<u | v> = \bar{u}_0 v_0 - \bar{u}_1 v_1 - \bar{u}_2 v_2$$

for $u = (u_0, u_1, u_2)$ and $v = (v_0, v_1, v_2)$; that is, J is the set of all x such that

$$^t(d\,x) = (\overline{dx})$$

where $d = \mathrm{diag}(1, -1, -1)$. The Cayley projective plane $P^2\mathbb{O}$ is defined as the subset I if non-zero elements x in J satisfying

(19.5) 1) $x^2 \lambda x$ with $\lambda \in R$.

2) x is not a sum of two non-zero elements satisfying 1).

Let modulo scalar multiplication; i.e., $P^2\mathbb{O} = I/R^x$, $\pi : I \to P^2\mathbb{O}$ denote the canonical map.

On J one introduces the bilinear form

$$(x, y) = \mathrm{tr}\, \frac{1}{2} (x \cdot y + y \cdot x) .$$

Set

$$I^+ = \{x \in I;\ (x, x) > 0\} . \tag{19.6}$$

A point x in I is nilpotent if and only if $(x, x) = 0$. One defines a Cayley *line* in $P^2\mathbb{O}$ to be a subset $\bar{x}$ of the form

$$\{\pi(u);\ u \in I,\ (x, y) = 0\} \tag{19.7}$$

where $x \in I$.

The elements in J have the form

$$x = \begin{pmatrix} u_0 & x_2 & \bar{x}_1 \\ -\bar{x}_2 & u_1 & x_0 \\ -x_1 & x_0 & u_2 \end{pmatrix}$$

and $x \in I$ if and only if all its 2×2 subdeterminants vanish.

We denote by X_0 the subset of $P^2\mathfrak{O}$ given by the nilpotent elements of I. The subset X_0 separates $P^2\mathfrak{O}$ into two connected components. A point x in $P^2\mathfrak{O}$ is called *inner* if every line containing x meets X_0; otherwise, x is called *outer*. One defines

(19.8) $\qquad H^2\mathfrak{O}$ is the set of inner points on $P^2\mathfrak{O}$.

Equivalently

$$H^2\mathfrak{O} = I^+/R^x \ . \tag{19.8'}$$

The group Aut J of automorphisms of J operates transitively on $H^2\mathfrak{O}$ and permutes transitively all the (Cayley) lines through points of $H^2\mathfrak{O}$.

We can determine the formula for an invariant metric on $H^2\mathfrak{O}$ by finding first the invariant metric on the line $\bar{u}$ where

$$u = \begin{pmatrix} 0 & 0 & 0 \\ 0 & 0 & 0 \\ 0 & 0 & 1 \end{pmatrix} .$$

The Cayley line u in $P^2\mathfrak{O}$ is given by the elements of the form

$$x = \begin{pmatrix} u_0 & x_2 & 0 \\ -\bar{x}_2 & u_1 & 0 \\ 0 & 0 & 0 \end{pmatrix}$$

with $u_0 u_1 + |x_2|^2 = 0$. The idempotent elements representing elements of $\bar{u} \cap H^2\mathfrak{O}$ satisfy the additional conditions

$$u_0 + u_1 = 1, \qquad u_0 \geq 1 \ .$$

Consequently,

$$\left(u_0 - \frac{1}{2}\right)^2 - |x_2|^2 = \frac{1}{4} \ .$$

Set $E_0 = \{x \in J;\ ux = 0,\ \mathrm{Tr}\ x = 0\}$. Then E_0 is the set of elements of the form

$$\begin{pmatrix} u & x_2 & 0 \\ -\bar{x}_2 & -u & 0 \\ 0 & 0 & 0 \end{pmatrix} \tag{19.9}$$

and $x \in I \cap \pi^{-1}(\bar{u})$ if and only if

$$x - \frac{1}{2}(e-u) \in E_0,\quad u^2 - |x_2|^2 = \frac{1}{4},\quad u \geq \frac{1}{2}$$

where e is the identity matrix, and $u = u_0 - \frac{1}{2}$. Clearly, E_0 is invariant under the stabilizer of u in Aut J. Conversely, by a result of T. A. Springer [18], any $\mathbf{R}$-linear norm-preserving automorphism of E_0 extends to an automorphism of the Jordan algebra J. Thus the stabilizer of u in Aut J is $SO(1,8)$ where E_0 is identified with $\mathbf{R}^9 = \mathbf{R} + \mathbf{O}$ by identifying the element (19.9) with (u, x_2). Any $SO(1,8)$ invariant metric on the quadric $u^2 - |x_2|^2 = \frac{1}{4}$ is, up to a constant factor,

$$\frac{1}{2}\cosh^{-1}\frac{\langle\langle x,y\rangle\rangle}{\langle\langle x,x\rangle\rangle^{\frac{1}{2}}\langle\langle y,y\rangle\rangle^{\frac{1}{2}}}$$

where $x = (u, x_2)$, $y = (v, y_2)$, and $\langle\langle x,y\rangle\rangle = uv - \mathrm{Re}\ x_2 y_2$. Translation by $-\frac{1}{2}(e-u)$ carries the idempotent elements in $I \cap \pi^{-1}(\bar{u})$ into E_0; defining the distance between points to be the distance between their translates in E_0, we find the formula

$$d(x,y) = 2^{-1}\cosh^{-1}\frac{2(x,y) - \mathrm{Tr}\,x\ \mathrm{Tr}\,y}{(2(x,x) - \mathrm{Tr}\,x)^2)^{\frac{1}{2}}(2(y,y)-(\mathrm{Tr}\,y)^2)^{\frac{1}{2}}} \tag{19.10}$$

$$= 2^{-1}\cosh^{-1}\frac{2(x,y) - \mathrm{Tr}\,x\ \mathrm{Tr}\,y}{\mathrm{Tr}\,x\ \mathrm{Tr}\,y}.$$

This formula is valid for any x and y representing points of $H^2\mathbf{O}$ because Aut J permutes transitively all points in $H^2\mathbf{O}$ and also all lines through points of $H^2\mathbf{O}$.

Note. The matrix u above defines an outer point of $P^2\mathbf{O}$ and its stabilizer is $SO(1,8)$. The result of Springer shows that the stabilizer of an inner point is Spin 9. Thus $H^2\mathbf{O} = F_4/\text{Spin } 9$.

These hyperbolic spaces constitute all the $\mathbf{R}$-rank 1 spaces. Let X be such a space, and let X_0 denote its boundary. Then X_0 is a sphere and for any $x_0 \in X_0$, its stabilizer G_{x_0} operates transitively on the complement of x_0 in X_0. The unipotent radical N of G_{x_0} also operates transitively on $X_0 - x_0$, and the Lie algebra of N can be described quite explicitly.

Let $d = \text{diag}(1,-1,\ldots,-1)$ denote the matrix of the hermitian form (19.1). Then the Lie algebra of its unitary group is given by the $(n+1) \times (n+1)$ matrices x with entries from $\mathbf{K}\,(\mathbf{K} = \mathbf{R},\ \mathbf{C}$ or $\mathbf{H})$ satisfying ${}^t\bar{x}d + dx = 0$. Thus x has the form

$$\begin{pmatrix} x_0 & \bar{x}_1 \ \cdots \ \bar{x}_n \\ x_1 & \\ \cdot & \\ \cdot & s \\ \cdot & \\ x_n & \end{pmatrix} \tag{19.11}$$

where s is an $n \times n$ skew-hermitian matrix, and $\bar{x}_0 = -x$. The isotropic vectors represent the points in X_0. Taking

$$v = \begin{pmatrix} 1 \\ 1 \\ 0 \\ \cdot \\ \cdot \\ \cdot \\ 0 \end{pmatrix}$$

the Lie algebra of its stabilizer consists of matrices

(19.12)
$$\begin{pmatrix} x_1 & -x_1 & \bar{x}_1 \dots \bar{x}_n \\ x_1 & -x_1 & \bar{x}_2 \dots \bar{x}_n \\ x_2 & -x_2 & \\ \cdot & \cdot & \\ \cdot & \cdot & s_1 \\ \cdot & \cdot & \\ x_n & -x_n & \end{pmatrix}$$

with s_1 an $(n-1) \times (n-1)$ skew-hermitian matrix and $x_1 = -x_1$. It follows that the Lie algebra $\mathfrak{N}$ of the unipotent radical of the stabilizer consists of matrices of the form (19.11) with $s_1 = 0$; such an element is denoted

$$(x_1;\, x)$$

where $x = (x_2, \dots, x_n)$ is in $\mathbf{K}^{n-1}$.

On $\mathbf{K}^{n-1}$ we define the inner product

$$(x, y) = \sum_2^n \bar{x}_2 y_2 \quad .$$

A direct calculation shows that if $(x_1; x)$ and $(y_1; y)$ are elements of $\mathfrak{N}$, then their Poisson bracket is given by $((x, y) - (y, x);\, 0)$. Thus

(19.13)
$$\mathfrak{N} \cong \mathbf{K}^{n-1} + \operatorname{Im} \mathbf{K}$$

where $\operatorname{Im} \mathbf{K} = \{c - \bar{c};\, c \in \mathbf{K}\}$ is central in $\mathfrak{N}$ and for any x, y in $\mathbf{K}^{n-1}$, the Poisson bracket is given by

(19.14)
$$[x, y] = 2 \operatorname{Im}(x, y) \; .$$

The formulas (19.12) and (19.13) are valid also for Cayley hyperbolic space $H^2_{\mathbf{O}}$; in this case $\dim \mathfrak{N} = 15$.

§20. The Boundary Semi-Metric

In the preceding section we have presented two models for hyperbolic space $H^n_{\mathbf{0}}$:

(i) The unit ball in $\mathbf{K}^n$ with metric (19.4) ($\mathbf{K} = \mathbf{R}, \mathbf{C}, \mathbf{H}$).

(ii) $H^2_{\mathbf{0}} = I^+/\mathbf{R}^x$ with metric (19.9) where I^+ denotes the set of all primitive idempotents x in the Jordan algebra J with $(x,x) > 0$.

The representation $I^+/\mathbf{R}^x$ was explicitly discussed only for $H^2_{\mathbf{0}}$ but actually it is valid for all $H^n_{\mathbf{K}}$.

In this section we wish to study the differentiability of certain set functions on the boundary of $H^n_{\mathbf{K}}$. For this purpose, model (i) is more convenient. We shall therefore modify model (i) above, so as to apply to $H^2_{\mathbf{0}}$ as well.

As in Section 19, let J be the Jordan algebra of all 3×3 matrices x with coefficients in θ satisfying ${}^t(dx) = \overline{dx}$, where $d = \text{diag}(1,-1,-1)$. On θ^2 we introduce the positive definite form

$$(v,w) = \bar{v}_1 w_1 + \bar{v}_2 w_2$$

and we set $D = \{v \in \mathbf{0}^2;\ (v,v) < 1\}$.

To get a set of representatives for $H^2_{\mathbf{0}}$ in J, we consider the subset I^1 of I whose entry in the (first row, first column) position is 1, then $I^{1+} = I^1 \cap I^+$ is a set of representatives for $H^2_{\mathbf{0}}$.

The projection of the elements in I^1 to their first column provides a bijective map of I^1 onto $\mathbf{0}^2$, whose inverse can be described as follows.

If $v = (v_1, v_2)$ is in $\mathbf{0}^2$, let v^1 denote the column vector

$$v^1 = \begin{pmatrix} 1 \\ v_1 \\ v_2 \end{pmatrix}$$

and set $j(v) = (v^1)\,{}^t\overline{(v^1)}d$: that is

$$(20.1) \qquad j(v) = \begin{pmatrix} 1 & -\overline{v_1} & -\overline{v_2} \\ v_1 & -|v_1|^2 & -v_1\overline{v_2} \\ v_2 & -v_2\overline{v_1} & -|v_2|^2 \end{pmatrix}.$$

Because any two elements in $\mathbf{0}$ generate an associative subalgebra, we find

$$(j(v))^2 = (v^1\,{}^t\overline{v}^{-1}d)(v^1\,{}^t\overline{v}^{-1}d) = ({}^t\overline{v}^{-1}d\,v^1)\,j(v)$$

where

$${}^tv^1\,d\,v^1 = 1-(v,v)\ .$$

Thus j maps $\mathbf{0}^2$ to I^1. The restriction of j to D gives a bijective map of D to I^{1+}.

Setting $x = j(v)$, $y = j(w)$, a direct calculation shows

$$(20.2) \quad (x,y) = |1-\bar{v}_1w_1-\bar{v}_2w_2|^2 + 2[\mathrm{Re}(v_1\bar{v}_2)(w_2\bar{w}_1)-\mathrm{Re}(\bar{v}_2w_2)(\bar{w}_1v_1)]\ .$$

Set

$$(20.3) \qquad R(v,w) = \mathrm{Re}(v_1\bar{v}_2)(w_2\bar{w}_1) - \mathrm{Re}(\bar{v}_2w_2)(\bar{w}_1v_1)$$

$$\langle v^1, w^1\rangle = 1-\bar{v}_1w_1-\bar{v}_2w_2$$

$$|v^1|^2 = \langle v^1,v^1\rangle = 1-|v|^2$$

$$|v|^2 = \langle v,v\rangle\ .$$

By (19.10) the distance between points in $H^2_{\mathbf{0}}$ is given by

$$d(x,y) = 2^{-1}\cosh^{-1}\frac{2(x,y)-\mathrm{Tr}\,x\ \mathrm{Tr}\,y}{\mathrm{Tr}\,x\ \mathrm{Tr}\,y}\ .$$

For $x = j(v)$, $y = j(w)$, this becomes $2^{-1}\cosh^{-1}(2c^2-1)$, where $c = (|\langle v^1,w^1\rangle|^2+2R(v,w))^{\frac{1}{2}}/|v^1|\cdot|w^1|$. Inasmuch as $\cosh^{-1}(2c^2-1) = 2\cosh^{-1}c$, we find

$$(20.4)\qquad d(x,y) = \cosh^{-1} \frac{(|1-(v,w)|^2 + 2R(v,w))^{\frac{1}{2}}}{(1-(v,v))^{\frac{1}{2}}\,(1-(w,w))^{\frac{1}{2}}}$$

exactly as in (19.4) for $H^n_{\mathbf{K}}$ with $\mathbf{K} \neq \mathbf{O}$. Thus we may take the right side of (20.4) as our definition of the metric in $H^n_{\mathbf{K}}$ in the unit ball model for all $\mathbf{K}$.

For any p, q, r in $\mathbf{O}$, we have the identity

$$\operatorname{Re} p(qr) = \operatorname{Re} q(rp) \ .$$

Thus we find

$$(20.5)\qquad R(v,w) = 0 \quad \text{if} \quad v_2 \in \mathbf{R} \ .$$

It is also clear that $R(v,w) = 0$ if v and w are in $\mathbf{H}^2$. Thus formula (20.4) may be regarded as an extension of formula (19.4) for the unit ball model of hyperbolic space.

REMARK. Any G invariant metric on $H^n_{\mathbf{K}}$ is unique up to a constant scalar. Our choice of metric satisfies the following normalizing identity:

$$(20.4')\qquad \log \alpha(a) = d(o, a(o))$$

for any polar element a stabilizing a ray with origin o, where α is the positive root on a for which $\alpha^{\frac{1}{2}}$ is not a root. In order to prove this, it suffices to check it in the case $H^2_{\mathbf{R}}$ for a single $a \neq 1$. In homogeneous coordinates, an element a with $\alpha(a) = \lambda$ is given by

$$x_0 + x_2 \to \lambda(x_0 + x_2)$$

$$x_0 - x_2 \to \lambda^{-1}(x_0 - x_2)$$

$$x_1 \to x_1 \ .$$

In non-homogeneous coordinates $(u_1, u_2) = (x_0^{-1}x_1, x_0^{-1}x_2)$, we find a sends the origin to $\tanh(\log \lambda)$. Formula (20.4) yields $d(o), a(o)) = \log \lambda$.

Let X denote H^n_K, realized as the open unit ball in K^n (where $n = 2$ if $K = O$) and let o denote the origin of K^n. We can regard K^n as a part of projective space P^n_K in the usual way for $K \neq O$, or via the map j of (20.1) even in the case $K = O$.

DEFINITION. The intersection of a K-line in P^n_K with H^n_K is called a K-line of H^n_K.

For $K = R$, C, or H the K-lines of K^n are merely the cyclic right K-submodules and their translates. However for $K = O$, this is not the case because of the non-associativity of multiplication.

LEMMA 20.1. *Let* $u = (u_1, u_2)$ *be a point in* O^2 *with* u_1 *or* $u_2 \in R$. *Then the* O*-line of* O^2 *through* u *and* o *is* uO.

Proof. Set $x = f(o)$ and $y = j(u)$. Then the K-line through x and y in the Jordan algebra model is given by the set of all points z in I^+ such that

$$(x \times y, z) = 0 \tag{20.6}$$

(cf. (19.7)), where $x \times y$, the "Freudenthal outer product" (cf. [5a]), is given by

$$x \times y = xy - \frac{1}{2}(y,e)x - \frac{1}{2}(x,e)y - \frac{1}{2}(x,y)e + \frac{1}{2}(x,e)(y,e)e$$

where xy is the product in J and e is the unit in J. For the sake of definiteness, we assume $u_1 \in R$. No generality is lost in assuming that $u_1^2 + |u_2|^2 = 1$. Thus $(y,e) = \mathrm{Tr}\, y = \mathrm{Tr}\, j(u) = 0$, and we find after a direct computation

$$x \times y = \frac{1}{2}\begin{pmatrix} 0 & 0 & 0 \\ 0 & -|u_2|^2 & u_1 \bar{u}_2 \\ 0 & u_1 u_2 & -|u_1|^2 \end{pmatrix}.$$

Let $z = j(v)$ with $v \in \mathbf{O}^2$. The condition that v lies on the $\mathbf{K}$-line through o and u is equivalent to the condition that z satisfy (20.6); that is

$$-|u_2|^2 v_1 + (u_1 \bar{u}_2) v_2 = 0 .$$

Since $u_1 \in \mathbf{R}$, this is equivalent to $v_2 = u_1^{-1} u_2 v_1$. The $\mathbf{K}$-line thus consists of the set of all points $(v_1, u_1^{-1} u_2 v_1)$ with $v_1 \in \mathbf{O}$; that is, $u\mathbf{O}$.

Note. Any $\mathbf{O}$-line in $P^2_{\mathbf{O}}$ is represented by an $\mathbf{R}$-linear subspace of I, and therefore projects onto an affine subspace via the projection of a matrix of $I^{1,+}$ onto its first column. Thus any $\mathbf{O}$-line in $H^2_{\mathbf{O}}$ lies on a real eight dimensional affine subspace of $\mathbf{O}^2$.

The $\mathbf{K}$-lines in $H^n_{\mathbf{K}}$ are geodesic subspaces, and they are permuted transitively by the group of isometries of $H^n_{\mathbf{K}}$. In addition to this family of geodesic subspaces, we shall also consider the family of real 2-planes.

DEFINITION. A real 2-plane is a geodesic two dimensional subspace of $H^n_{\mathbf{K}}$ lying in no $\mathbf{K}$-line.

The geodesic rays through o are given by the 1 dimensional $\mathbf{R}$-subspaces of $\mathbf{K}^n$ and the real 2-planes through o are given by pairs of vectors u and v in $\mathbf{K}^n$ such that the $\mathbf{K}$-line through o containing u is orthogonal to the $\mathbf{K}$-line through o containing v. The real 2-planes through o are permuted transitively by G_o, the stabilizer of o. For the case $H^2_{\mathbf{O}}$, a word of explanation is in order. Here $G_o = \text{Spin } 9$ operating on $\mathbf{O}^2 = \mathbf{R}^{16}$ via the spinor representation. The stabilizer of the $\mathbf{K}$-line through $(0,0)$ and $(1,0)$ is Spin 8 and the stabilizer of the geodesic ray R from $(0,0)$ to $(1,0)$ is Spin 7. Let $L_1 = \mathbf{O} \times 0$ and $L_2 = 0 \times \mathbf{O}$ denote the coordinate $\mathbf{O}$-axes in $\mathbf{O}^2$. Then G_{L_1} operates on L_1 as SO(8) via the even $\frac{1}{2}$-spin representation, but its operation on L_2 is given by the odd $\frac{1}{2}$-spin representation. Thus the stabilizer G_R of the ray R operates on L_2 via the spin representation of Spin 7. In particular, G_R is transitive on all the geodesic rays through o in L_2. It follows at once that G_o is transitive on pairs of vectors $\{u, v\}$ with $u \in L_1$, $v \in L_2$ and $|u| = |v| = 1$. In particular, G_o is transitive on the family of real 2-planes through o.

Let X_0 denote the boundary of the $\mathbf{R}$-rank one space X. In our model, X_0 is the boundary sphere S^{kn-1}, where $k = \dim_{\mathbf{R}} \mathbf{K}$.

DEFINITION. A *great* $\mathbf{K}$-*sphere* is the boundary in X_0 of a $\mathbf{K}$-line in X through o. A *great* $\mathbf{R}$-circle is the boundary in X_0 of a real 2-plane in X through o. More generally, the boundary in X_0 of a $\mathbf{K}$-line (resp. real 2-plane) in X is called a $\mathbf{K}$-sphere (resp. $\mathbf{R}$-circle).

Clearly, a great $\mathbf{K}$-sphere is a (—1 dimensional sphere which is the image under an element of G_o, the stabilizer of o, of the unit sphere S^{k-1} in the first coordinate axis. The boundary $\mathbf{K}$-spheres make up the Hopf fibering of S^{kn-1} by S^{k-1} $(k = 1, 2, 4, 8)$.

It is easy to see that each great $\mathbf{R}$-circle cuts each great $\mathbf{K}$-sphere orthogonally with respect to the standard spherical metric.

The stabilizer G_o also permutes transitively all the great $\mathbf{R}$-circles. Furthermore our remarks above on the stabilizer of a ray through o in $H^2_{\mathbf{O}}$ shows that in $H^2_{\mathbf{O}}$ as well as in any $H^n_{\mathbf{K}}$, we have:

(20.7). For any $p \in X_0$, the stabilizer G_{op} of p and o permutes transitively the set of all great $\mathbf{R}$-circles through p, and it stabilizes the great $\mathbf{K}$-sphere through p.

Another observation that will be useful is

(20.8). Given any two great $\mathbf{R}$-circles S_1 and S_2, and given three distinct points p, q, r in S_1, and given that the great $\mathbf{K}$-sphere through p meets S_2 and the great $\mathbf{K}$-sphere through q meets S_2, then the great $\mathbf{K}$-sphere through r meets S_2.

Proof. Without loss of generality we can assume that $p = (1, 0, \ldots, 0)$ and S_1 consists of the points $(t, (1-t^2)^{\frac{1}{2}}, 0 \ldots 0)$ with $-1 \leq t \leq 1$ then the great $\mathbf{K}$-sphere through p consists of the points $(k, 0, \ldots 0)$ with $k \in \mathbf{K}$ and $|k| = 1$. Thus S_2 contains points $u = (ta, (1-t^2)^{\frac{1}{2}}a, 0, \ldots, 0)$ and $v = (b, 0, \ldots, 0)$ with a and b in $\mathbf{K}$ and $|a| = |b| = 1$. The fact that S_2 is the boundary of a real 2-plane through o implies that $(u, v) \in \mathbf{R}$ by (20.5). Hence $a = \pm b$. It follows at once that $ra \in S_2$. By Lemma 20.1, this implies that the great $\mathbf{K}$-sphere through r meets S_2.

REMARK. The term "$\mathbf{R}$-circle" is prone to misunderstanding because not every $\mathbf{R}$-circle is in fact a circle with respect to the standard Euclidean metric in $\mathbf{K}^n$. A sufficient condition that the boundary of a real 2-plane P be a circle with respect to the Euclidean metric is that P lie in a two dimensional $\mathbf{R}$-affine subspace of $\mathbf{K}$. (This is always the case if $\mathbf{K} = \mathbf{R}$ but not always if $\mathbf{K} \neq \mathbf{R}$.) By this criterion, a great $\mathbf{R}$-circle is a Euclidean circle. On the other hand a $\mathbf{K}$-sphere is always a Euclidean sphere, since any $\mathbf{K}$-line lies in a k-dimensional $\mathbf{R}$-affine subspace of $\mathbf{K}^n$ (cf. Note following Lemma 20.1).

Let L denote the $\mathbf{K}$-line in X, given by the points in $\mathbf{K}^n$ whose last $n-1$ coordinates are zero, and let S^{k-1} denote the boundary of L. Let $S^{k(n-1)-1}$ denote the boundary of the hyperplane of points in X whose first coordinate is zero. Each pair of points $(p, q) \in S^{k-1} \times S^{k(n-1)-1}$ determines a unique great $\mathbf{R}$-circle; with respect to the standard metric on the sphere X_0, each such great $\mathbf{R}$-circle is orthogonal to both S^{k-1} and $S^{k(n-1)-1}$. Let C denote the union of all $\mathbf{R}$-circles through $(1, \ldots, 0)$ and points on $S^{k(n-1)-1}$ with the latter points deleted; set $X_0^+ = X_0 - S^{k(n-1)-1}$. The transforms of C under the stabilizer G_L of the $\mathbf{K}$-line L provide a homogeneous fibering π of X_0^+. We set

$$C_p = \text{the fiber of } \pi \text{ through the point } p \,. \tag{20.11}$$

Set $X_* = X_0 - S^{k(n-1)-1} - S^{k-1}$; that is, the set of points of the boundary that lie in neither the first coordinate axis of $\mathbf{K}^n$ nor in its orthogonal complement. On X_* we shall introduce two fiberings, $\pi_{\mathbf{R}}$

and $\pi^{\mathbf{K}}$. Let $\pi_{\mathbf{R}}$ denote the fibering of X_* having as typical fiber the connected component of a great $\mathbf{R}$-circle passing through S^{k-1} and $S^{k(n-1)-1}$. Let $\pi^{\mathbf{K}}$ denote the restriction to X_* of the Hopf fibering of X_0.

By virtue of (20.8), the Hopf fibering $\pi^{\mathbf{K}}$ induces a well-defined fibering on the base space $\pi_{\mathbf{R}}(X_*)$; we shall denote this induced fibering of $\pi_{\mathbf{R}}(X_*)$ by the same symbol $\pi^{\mathbf{K}}$. Thus we have a commutative diagram

$$\begin{array}{ccc} X_* & \xrightarrow{\pi_{\mathbf{R}}} & \pi_{\mathbf{R}}(X_*) \\ \pi^{\mathbf{K}}\downarrow & & \downarrow\pi^{\mathbf{K}} \\ \pi^{\mathbf{K}}(X_*) & \longrightarrow & \pi^{\mathbf{K}}(\pi_{\mathbf{R}}(X_*)) \end{array} \tag{20.12}$$

Set $Y = \pi_{\mathbf{R}}(X_*)$. We introduce a metric on Y as follows. The boundary of the tube $T_{\frac{1}{2}}(S^{k-1})$ of radius $\frac{1}{2}$ around the first coordinate $\mathbf{K}$-sphere S^{k-1} meets each fiber of $\pi_{\mathbf{R}}$ in a single point, thus providing a continuous cross-section $\psi: Y \to X_*$ for the fibering $\pi_{\mathbf{R}}$. For any $(y_1, y_2) \in Y \times Y$, set

$$\delta(y_1, y_2) = |\psi(y_1) - \psi(y_2)| \tag{20.13}$$

where $|p-q|$ denotes the standard Euclidean distance on $\mathbf{K}^n$.

We now introduce a function $d^{\mathbf{K}}$ on $\mathbf{K}^n \times \mathbf{K}^n$ which on the one hand is related to the hyperbolic metric on $H^n_{\mathbf{K}}$ and on the other hand underlies the notion of a *quasi-conformal mapping over* $\mathbf{K}$. Define $d^{\mathbf{K}}$ by the formula

$$\begin{aligned} 4^{-1}d^{\mathbf{K}}(v,w)^4 &= |1-(v,w)|^2 + 2R(v,w) - (1-|v|^2)(1-|w|^2) \\ &= |v-w|^2 + 2R(v,w) - (|v|^2|w|^2 - |(v,w)|^2) \end{aligned} \tag{20.14}$$

where $(v,w) = \sum_1^n \bar{v}_i w_i$ for $v = (v_1, \ldots, v_n)$ and $w = (w_1, \ldots, w_n)$, and where $R(v,w)$ is given by (20.3) if $\mathbf{K} = \mathbf{O}$, $n = 2$ but is zero in all other cases. By (20.2), the right hand side of (20.14) is $(x,y) - \operatorname{Tr} x \operatorname{Tr} y$ where

$x = j(v)$ and $y = j(w)$ and thus it is invariant under the stabilizer G_o of the origin o. For any $v \in \mathbf{K}^n$ and $r > 0$, set

$$B^{\mathbf{K}}(v,r) = \{w \in \mathbf{K}^n;\ d^{\mathbf{K}}(v,w) \leq r\} ;$$

we call this set the **K**-ball of center v and radius r.

REMARK 1. A **K**-ball with center at the point v and radius $(2s)^{\frac{1}{2}}$, where $|v| < 1$, is an ellipsoid with respect to the standard Euclidean metric on $\mathbf{K}^n$ whose intersection with the **K**-line through the origin and v is a standard ball of center v and radius s, and whose intersection with any real 2-plane through the origin and v is an ellipse of semi-major axis $s/(1-|v|^2)^{\frac{1}{2}}$ and semi-minor axis s along the ray ov. In order to verify this, it suffices to take $v = (a, 0, \ldots, 0)$ with a real – for the stabilizer G_o is transitive on rays through the origin. In this situation, $R(v,w) = 0$ for all w in $\mathbf{K}^n$ and the equation of the boundary of $B^{\mathbf{K}}(v, \sqrt{2s})$ becomes

$$(20.14)' \qquad |v-w|^2 - a^2\,|w|^2 + a^2\,|w_1|^2 = s^2$$

where $w = (w_1, w_2, \ldots.)$; or alternatively

$$(20.14)'' \qquad |a-w_1|^2 + (1-a^2)(|w_2|^2 + \ldots + |w_n|^2) = s^2$$

which is clearly an ellipsoid. If w lies on the **K**-line through v and o, then $w = (w_1, 0, \ldots, 0)$ and (20.14)″ becomes

$$|a - w_1| = s\ .$$

If w lies in a real 2-plane through v and o, then we can assume without loss of generality that $w = (w_1, w_2, 0, \ldots, 0)$ with w_1 and w_2 real and (20.14)″ becomes

$$(w_1-a)^2 + (1-a^2)\,w_2^2 = s^2\ .$$

This proves our remark.

REMARK 2. If $|v| = 1$, then in the standard metric on $\mathbf{K}^n$, $B^{\mathbf{K}}(v,(2s)^{\frac{1}{2}})$ becomes the unbounded cylinder whose base is a standard ball of radius s in the K-line through the origin and v. To see this, we can assume without loss of generality that $v = (1, 0, \dots, 0)$. Then setting $a = 1$ in (20.14)″, we get

$$(20.14)''' \qquad |w_1 - a| = s$$

as the equation of the boundary of the cylinder.

For any v in K^n with $|v| < 1$ and for any $r > 0$, set

$$B(v,r) = \{w \in H^n_K;\ d(v,w) \leq r\} ;$$

that is, the ball in H^n_K of center v and radius r. We shall see below in Section 21 that every $B(v,r)$ of radius bounded away from zero is, up to bounded distortion, a K-ball and conversely, every K-ball lying in the unit ball close to the boundary X_0 is, up to bounded distortion, a ball in the hyperbolic metric (cf. (21.4) and (21.6)).

REMARK 3. It is easy to see (cf. (21.4) and (21.6)) that the ball $B(v,r)$ is an ellipsoid in the standard Euclidean metric on K^n, even when $K = R$. This does not conflict with the fact that hyperbolic balls are Euclidean balls in the Poincaré model of H^n_R. One can pass from our model of H^n_R to the well-known Poincaré model by first projecting the unit ball in R^n onto the upper hemisphere of the unit sphere and then applying stereographic projection (cf. [12g]).

Let d_0 denote the restriction of d^K to $X_0 \times X_0$. For any $p \in X_0$ and $s > 0$, set

$$(20.15) \qquad K(p,s) = B^K(p,s) \cap X_0 .$$

We call $K(p,s)$ the d_0-*ball* of center p and radius s; we also refer to it as a "*boundary-ball.*"

The next lemma shows that if we attempted to form a metric on X_0 from d_0, it would yield the usual geodesic arc length along great R-circles but would yield infinite arc length along great K-spheres. For this reason we call d_0 a *semi-metric*.

LEMMA 20.2.

(i) *For any* p,q *on a great* $\mathbf{K}$*-sphere*, $d_0(p,q) = (2|p-q|)^{\frac{1}{2}}$

(ii) *For any* p,q *on a great* $\mathbf{R}$*-circle*, $d_0(p,q) = |p-q|$

(iii) *For any* p,q *on* X_0, $d_0(p,q) \geq |p-q|$.

Proof. Let G_o denote the stabilizer of the origin o. Since G_o preserves both d_0 and the standard Euclidean distance $|p-q|$ on $\mathbf{K}^n$, and since G_o operates transitively on X_0, no generality is lost in taking $p = (1,\ldots,0)$. Then $R(p,q) = 0$ for all $q \in \mathbf{K}^n$ by (20.5). The $\mathbf{K}$-sphere through p consists of points $q = pa$ with $a \in \mathbf{K}$ and $|a| = 1$. Thus $d_0(p,q)^2 = 2|1-a| = 2|p-q|$. This proves (i).

To prove (ii), we can assume that $p = (1,\ldots,0)$ and $q = (a,b,0,\ldots,0)$ with $(a,b) \in \mathbf{R}^2$ and $a^2+b^2 = 1$, by (20.7). Then $|p-q| = |(1-a,b,0,\ldots,0)|$ $= ((1-a)^2+b^2)^{\frac{1}{2}} = (2-2a)^{\frac{1}{2}} = 2^{\frac{1}{2}}(1-(p,q))^{\frac{1}{2}} = d_0(p,q)$.

To prove (iii), we can assume that $p = (1,\ldots,0)$ and $q = ap + bp^{\perp}$ with $(p,p^{\perp}) = 0$, $|p^{\perp}| = 1$, and $|a|^2+|b|^2 = 1$, then

$$d_0(p,q)^2 = 2|1-a| \geq 2(1-\operatorname{Re} a) = |1-a|^2+|b|^2 = |p-q|^2 .$$

The set functions that we shall consider will be defined on the base space Y of the fibering $\pi_{\mathbf{R}}$ described above. Before discussing the differentiability of our set functions, we require some auxiliary definitions.

Let $p_0 = (1,0,\ldots,0)$ and let $0 < s < 1$. By a polydisc $D_0(p_0,s)$ of center p_0 and radius s we mean the subset

$$D_0(p_0,s) = \{w \in X_0;\ |w_{\mathbf{K}}-p_0| \leq s^2/2,\ |w_{\mathbf{K}}-w| \leq s\}$$

where $w_{\mathbf{K}} = C_w \cap S^{k-1}$, and C_w is the fiber of π through w (cf. (20.11)). Given any $p \in X_0$, and any $g \in G_o$ such that $gp_0 = p$, we define the polydisc

$$D_0(p,s) = gD_0(p_0,s) .$$

Inasmuch as G_o is transitive on X_0 and the stabilizer $G_o \cap G_{p_0}$ stabilizes $D(p_0,s)$, the polydiscs at any point of X_0 are well-defined. The notion of a polydisc on the base space $Y = \pi_R(X_*)$ is defined by means of the cross-section $\psi : Y \to X_*$; namely for any $y \in Y$ and s with $0 < s < 1$ we set

(20.16) $$D(y,s) = \pi_R \, D_0(\psi(y),s) \ .$$

LEMMA 20.3. *Let Φ be a completely additive finite-valued set function defined on all closed subsets of the space $Y = \pi_R(X_*)$. Then for almost all $y \in Y$,*

$$\lim_{s \to 0} \sup \frac{\Phi(D(y,s))}{\mu(D(y,s))} < \infty$$

where $\mu(\ )$ denotes $kn-2$ dimensional measure in Y.

Proof. This lemma asserts the analogue of Lebesgue's theorem on differentiation of set functions with polydiscs taking the role of balls; this yields of course a modified notion of a family of sets with a "modulus of regularity." The proof of Lebesgue's differentiation theorem depends on the validity of Vitali's covering theorem for the family of closed sets with respect to which one differentiates. In our case, it is a simple matter to verify that the usual proof of the Vitali covering theorem holds, mutatis mutandis, for the family of polydiscs on Y. (cf. for example, E. J. McShane "Integration," Princeton University Press, pp. 366-9.)

We come now to an important comment relating d_0-balls in X_0 and polydiscs in Y. Let U denote the open tube of positive radius u (in the standard Euclidean metric of K^n) about $X_0 - X_*$, and set $X_1 = X_0 - U$. Then

(20.17) There is a positive constant c such that for any $p \in X_1$ and $s > 0$

$$D(\pi_R(p), c^{-1}s) \subset \pi_R(K(p,s)) \subset D(\pi_R(p), c\,s) \ .$$

To prove (20.17), it clearly suffices to prove the assertion for all small s.

Set $p_0 = (1,0,\dots,0)$. It follows at once from definitions that $B^{\mathbf{K}}(p_0,s)$ is the set of points $w = (w_1,\dots,w_n)$ with $|w_1-1| \le s^2/2$ (cf. (20.14)‴). Therefore for any w in $\mathbf{K}(p_0,s) = B^{\mathbf{K}}(p_0,s) \cap X_0$, we have $|w-1|^2 = |w_1-1|^2 + |w_2|^2 + \dots + |w_n|^2 \equiv |w_1|^2 - 2 \operatorname{Re} w_1 + 1 + 1 - |w_1|^2 = 2(1-\operatorname{Re} w_1) \le 2|1-w_1| \le s^2$. Thus

$$\mathbf{K}(p_0,s) \subset D_0(p_0 s) .$$

For small s, $\mathbf{K}(p_0,s)$ is approximately an ellipsoid of semi-major axis s and semi-minor axis $s^2/2$, and one finds that for all small s,

$$D_0(p_0,s/2) \subset \mathbf{K}(p_0,s) \subset D_0(p_0,s) .$$

It follows that for all $p \in X_0$ and small s,

$$D_0(p,s/2) \subset \mathbf{K}(p,s) \subset D_0(p,s) .$$

Next we observe that there is a positive constant c such that for any $p \in X_1$,

$$D(\pi_{\mathbf{R}}(p),c^{-1}s) \subset \pi_{\mathbf{R}} D_0(p,s) \subset D(\pi_{\mathbf{R}}(p),c\,s) .$$

This follows from the compactness of the fibers in X_1 together with the fact that for small s, $\mathbf{K}(p,s)$ is approximately an ellipsoid with a principal axis lying along the fiber of $\pi_{\mathbf{R}}$. From these observations, (20.17) follows.

Note. In case $\mathbf{K} = \mathbf{R}$, $\mathbf{K}(p,s) = D_0(p,s)$ is a standard ball in the unit sphere of *chordal* radius s.

We conclude this section with a simple observation about the action of the stabilizer G_p of a point $p \in X_0$ on the tangent space to X_0 at p. Let N denote the unipotent radical of G_p and let G_{op} denote the stabilizer of the origin o and the point p. Then

$$G_p = G_{op} A N$$

where $A = G_p \cap G_{-p}$ and is a maximal polar subgroup of G. Let

$$\dot{G} = Z(\dot{A}) + \dot{G}_{\alpha} + \dot{G}_{2\alpha} + \dot{G}_{-\alpha} + \dot{G}_{-2\alpha}$$

be the R-root space decomposition of the Lie algebra $\dot{G}$ with respect to A. Then $\dot{G}_{\alpha} + \dot{G}_{2\alpha}$ and $\dot{G}_{-\alpha} + \dot{G}_{-2\alpha}$ are precisely $\dot{N}$ and $\dot{N}^{-}$, where N^{-} denotes the unipotent radical of G_{-p}. Assume $\dot{N} = \dot{G}_{\alpha} + \dot{G}_{2\alpha}$. We can identify the tangent space to X_0 at p with $\dot{G}/\dot{G}_p$. The root space relations $[\dot{G}_{\beta}, \dot{G}_{\gamma}] \subset \dot{G}_{\beta+\gamma}$ yield $[\dot{N}, \dot{G}_{-\alpha}] \subset \dot{G}_p$ and $[\dot{N}, \dot{G}_{-2\alpha}] \subset \dot{G}_{-\alpha} + \dot{G}_p$. From this we infer at once that N leaves fixed each tangent vector in $(\dot{G}_{-\alpha} + \dot{G}_p)/\dot{G}_p$ and shears each vector in $(\dot{G}_{-2\alpha} + \dot{G}_p)/\dot{G}_p$ along $(\dot{G}_{-\alpha} + \dot{G}_p)/\dot{G}_p$.

It is easy to see that $(\dot{G}_{-\alpha} + \dot{G}_p)/\dot{G}_p$ is the set of tangent vectors to the great R-circles through p and that $(\dot{G}_{-2\alpha} + \dot{G}_p)/\dot{G}_p$ is the tangent space to the great K-sphere through p. Thus we conclude

(20.18) *The unipotent group* N *keeps fixed each tangent vector to a great* R-*circle at* p, *and shears each tangent vector to the* K-*sphere at* p *along the tangent to a great* R-*circle through* p.

It is easy to see that any R-circle is uniquely determined by three of its points, and that G_p permutes transitively all the R-circles through p. Since $G_{op}A(= Z(A))$ stabilizes $\dot{G}_{-\alpha}$, we see from (20.18) that G_p stabilizes the subspace $(\dot{G}_{-\alpha} + \dot{G}_p)/\dot{G}_p$ of the tangent space to X_0 at p; we denote this subspace by X_p^R. This (n–1)k-dimensional vector-space is the set of all tangent vectors to R-circles at p. The assignment $p \to X_p^R$ is an (n–1)k-contact distribution on X_0; it is not involutive if $k > 1$. We have

(20.19) The distribution $p \to X_p^R$ is G-invariant.

§21. Quasi-Conformal Mappings Over K and Absolute Continuity on Almost All R-Circles

Let X and X' be R-rank 1 spaces as in Section 19, and let G and G' denote their respective groups of isometries. Let Γ and Γ' be discrete subgroups of G and G' respectively, and let $\theta:\Gamma \to \Gamma'$ be an isomorphism. We *assume*

(i) G/Γ *and* G'/Γ' *have finite Haar measure.*

(ii) *There exists pseudo-isometries* $\phi: X \to X'$ *and* $\phi': X' \to X$ *which are* Γ*-space morphisms.*

By Theorem 14.2, these conditions are satisfied if G/Γ and G'/Γ' are compact. By Theorem 15.2 the map ϕ induces a boundary map $\phi_0: X_0 \to X'_0$ which is a homeomorphism. We shall show that ϕ_0 is differentiable on X_0. Our proof will come in several steps.

First we relate the boundary balls on X_0 to the balls of the space X. We use the model for X described in Section 20. In particular, o denotes the origin of X, and π_R denotes the fibration of $X_* = X_0 - S^{k(n-1)-1} - S^{k-1}$ with fibers that lie along great R-circles.

LEMMA 21.1. *Let* ∞ *be a point on the boundary* X_0 *and let* $\rho: X \to X_0$ *denote the map:* $\rho(v) = p$, *where* p *is the second endpoint of the infinite geodesic line from* ∞ *through* v.

Set

$$B(v,r) = \{w \in X;\ d_X(v,w) \leq r\}$$

where d_X *denotes the hyperbolic metric on* X. *Let* $p \in X_0$ *and assume* $\rho(v) = p$. *Set*

$$R = d_X(o,v)\ ,$$

and assume that the ray from v *to* p *lies at a distance* R *from* p. *Then for all* R *sufficiently large*

(21.1) $$K(p,5^{-1}e^{-R+r}) \subset \rho(B(v,r)) \subset K(p,2^{-1}e^{-R+r}) .$$

Proof. The geodesic lines in the unit ball model for $\mathbf{K}^n$ are easy to describe. Each geodesic line F lies in a unique **K**-line L which is a k-dimensional geodesic subspace, $k = \dim_{\mathbf{R}} \mathbf{K}$. In turn, the **K**-line L lies on a k-dimensional **R**-affine subspace of $\mathbf{K}^n$ and meets the unit ball X in a k-dimensional disc. Furthermore, this geodesic subspace is an $H^k_{\mathbf{R}}$ in which the geodesics are the arcs of circles meeting the boundary of the disc orthogonally. In the case $\mathbf{K} = \mathbf{R}$, the geodesic lines thus lie on straight lines of $\mathbf{R}^n$.

Inasmuch as the group G of isometries of X is transitive on X_0, no generality is lost in assuming that $p = (1,0,...)$; that is, we assume that the last $n-1$ coordinates of p are zero. Set $s = |v|$, $\sigma = 1-s$, and $h = \cosh r$.

We consider first the special case that $\infty = (-1,0,...)$. Then $v = (|v|,0,...) = (s,0,...)$. By (20.4) and (20.3), the term $R(v,w)$ of the metric formula vanishes and

(21.2) $$d_X(v,w) = \cosh^{-1} \frac{|1-(v,w)|}{(1-|v|^2)^{\frac{1}{2}}(1-|w^2|)^{\frac{1}{2}}}$$

for any $w \in X$. Thus the boundary of the ball $B(v,w)$ is given by the quadratic equation in the kn real coordinates of the point $w \in \mathbf{K}^n$:

(21.3) $$|1-(v,w)|^2 = h^2(1-s^2)(1-|w|^2) .$$

We see therefore that $B(v,r)$ is an ellipsoid whose principal axes are parallel to the coordinates planes of $\mathbf{K}^n$. Similarly, we see that the d_0-ball $K(p,t)$ is the intersection of X_0 with an ellipsoid in $\mathbf{K}^n$ centered at p whose principal axes are parallel to those of $B(v,r)$. On the other hand, up to infinitesimals of higher order, the map ρ is the projection of $B(v,r)$ along straight lines in $\mathbf{K}^n$ parallel to the diameter ∞_p. In order to prove (21.1), it suffices to compare the image of the principal axes of $B(v,r)$ under ρ with those of $K(p,t)$.

Let G_{op} denote the subgroup of G stabilizing o and p. Inasmuch as G_{op} operates transitively on the orthogonal complement to op in the K-sphere $pK \cap X_0$ and also on the orthogonal complement $S^{k(n-1)-1}$ to pK in X_0, no generality is lost in considering just two types of principal axes of $B(v,r)$:

Type I, lying in the C-line pC.

Type II, lying in the real 2-plane $(x,y,0\dots 0)$ with $(x,y) \in R^2$.

Consider first Type I. Here $w = (x+iy, 0, \dots)$. Writing $|1-sw|^2 = (1-sx)^2 + s^2y^2$, we get from (21.3)

$$(h^2 - s^2(h^2-1))(x^2+y^2) - 2sx = h^2(1-s^2) - 1$$

or

$$\left(x - \frac{s}{h^2-s^2(h^2-1)}\right)^2 + y^2 = h^2(h^2-1)(1-s^2)^2 \tag{21.4}$$

since $h^2(h^2-1)(1+s^4) - (h^4+(h^2-1)^2)s^2 + s^2 = h^2(h^2-1)(1-s^2)^2$. As s approaches 1, we have $1-s^2 = (1-s)(1+s) \approx 2\sigma$ and $h^2-s^2(h^2-1) \approx 1 + 2(h^2-1)\sigma$; the approximation

$$a \approx b \tag{21.5}$$

means for us that $a/b \to 1$ as $\sigma \to 0$. Thus (21.4) is the equation of circle in the complex line pC with center at approximately $1-(2h^2-1)\sigma$ and radius approximately $2h(h^2-1)^{\frac{1}{2}}\sigma$. For r fixed and large R, ρ is given, up to infinitesimals of higher order, by the projection along straight lines parallel to ∞p. Thus ρ maps the circle (21.4) onto an arc in X_0 of semi-length approximately $2h(h^2-1)^{\frac{1}{2}}\sigma = (\sinh 2r)\sigma$.

To compute the effect of ρ on the second type of principal axes, we take $w = (x,y,0\dots)$ on the intersection of the boundary of $B(v,r)$ with the real 2-plane. Since $|1-(v,w)|^2 = |1-sx|^2$, we get from (21.3)

$$(h^2-s^2(h^2-1))x^2 + h^2(1-s^2)y^2 - 2sx = h^2(1-s^2) - 1 \ .$$

Setting $\sigma = 1-s$, we get as $\sigma \to 0$

$$(21.6) \qquad \left(x - \frac{s}{h^2-s^2(h^2-1)}\right)^2 + 2h^2\,\sigma y^2 \quad 4h^2(h^2-1)\sigma^2 \ .$$

This is an ellipse with center at approximately $1-(2h^2-1)\sigma$ and semi-major axis perpendicular to op of length $(2(h^2-1)\sigma)^{\frac{1}{2}}$, which is $(2\sigma)^{\frac{1}{2}}$ sinh r. The map ρ sends this major axis onto an arc in X_0 of approximately the same length. The semi-minor axis of the ellipse (21.6) is the same as the radius of the circle (21.4) $2h(h^2-1)^{\frac{1}{2}}\sigma$.

By Lemma 20.2, the d_0-ball $\mathbf{K}(p,t)$ has a semi-major axis of length t in the approximate sense of (21.5) along any $\mathbf{R}$-circle through p; and if $\mathbf{K} \neq \mathbf{R}$, $\mathbf{K}(p,t)$ has a semi-minor axis of length $t^2/2$ approximately along any $\mathbf{K}$-sphere. Thus $\mathbf{K}(p,(2(h^2-1)\sigma)^{\frac{1}{2}}(2h/(h^2-1)^{\frac{1}{2}})^{\frac{1}{2}})$ has the same minor axis as $\rho(B(v,r))$ approximately, if $\mathbf{K} \neq \mathbf{R}$, and a larger major axis each along matching directions. Similarly, $\mathbf{K}(p,(2(h^2-1)\sigma)^{\frac{1}{2}})$ has approximately the same major axis as $\rho(B(v,r))$ but a smaller minor axis.

From (21.2) we see that $R = \cosh^{-1}(1-s^2)^{-\frac{1}{2}}$. Thus $\cosh^2 R = (1-s^2)^{-1} \approx (2\sigma)^{-1}$ and $(2\sigma)^{\frac{1}{2}} \approx 2^{-1}e^{-R}$. Moreover $(h^2-1)^{\frac{1}{2}} = \sinh r$. Thus for R sufficiently large and for $r > 2^{-1}\log 3$, we find

$$(21.7) \qquad \begin{gathered} 5^{-1}\,e^{-R+r} < (2(h^2-1)\sigma)^{\frac{1}{2}} \\ (2(h^2-1)\sigma)^{\frac{1}{2}}\left(2h/(h^2-1)^{\frac{1}{2}}\right)^{\frac{1}{2}} < 2^{-1}\,e^{-R+r} \ . \end{gathered}$$

From these inequalities and the remarks above, we infer (21.1). This completes the proof for the special case that $\infty = (-1,0,\ldots)$.

Suppose now that ∞ is any point on X other than p. Let $q = (-1,0,\ldots)$, and let ρ_q denote the projection of X into X_0 along geodesic lines issuing from q; that is ρ_q is the map described in the lemma for $\infty = q$. Let $\blacktriangleleft F$ denote the ray op, and let F denote the line qp. Let A denote the maximal polar subgroup of the stabilizer G_F and let N denote the unipotent radical of G_p. Then A operates simply transitively on F and N operates simply transitively on X_0-p. Let g be the element of N such that $g(q) = \infty$. Then

$$\text{(21.8)} \qquad \rho = g\,\rho_q\,g^{-1}$$

since the isometry g carries geodesics to geodesics. Let a denote the element of A such that $a(o) = v$. Then for any $x \in B(v,r)$ we have $x = a(y)$ with $y \in B(p,r)$. Thus $d_X(x,gx) = d_X(a(y), g^{-1}a(y)) = d(y,a^{-1}g^{-1}a(y)) < ce^{-R}$ for some constant c depending on g; for by our choice of metric in X (cf. (20.4′)) $d(o,a(o)) = \log \alpha(a)$ if $\alpha(a)$ is the reduced positive root on a. Therefore, $g^{-1}(B(v,r)) \subset B(v,r^+ce^{-R})$. On the other hand, the differential of g at the tangent space to X_0 at p is the identity along the great $\mathbf{R}$-circles at p and shears each tangent vector to the great $\mathbf{K}$-sphere at p along a tangent vector to a great $\mathbf{R}$-circle at p (cf. (20.18)). Therefore, up to infinitesimals of higher order, each point of $\mathbf{K}(p,t)$ is moved by g a distance at most $c't^2$ along a direction parallel to a great $\mathbf{R}$-circle, when t is small, c' being a constant depending on g. Thus

$$\text{(21.9)} \qquad g(\mathbf{K}(p,t)) \approx \mathbf{K}(p,t) \quad \text{as} \quad t \to 0\ .$$

Combining these approximations and the validity of (21.10) in the case $q = \infty$, we infer the validity of (21.10) for any $q \in X_0 - p$.

The next lemma asserts the key property of the boundary map ϕ_0 which opens the door to analyzing its differentiability properties. It states in effect that the map ϕ_0 is K-quasi-conformal with respect to the semi-metric d_0.

LEMMA 21.2. *There is a positive constant* κ *such that: For all* $p \in X_0$ *and for all sufficiently small* $t > 0$, *there is an* $s > 0$ *such that*

$$K'(\phi_0(p),s) \subset \phi_0(\mathbf{K}(p,t)) \subset \mathbf{K}'(\phi_0(p),\kappa s) \ . \tag{21.10}$$

Proof. To be more explicit, we have $X = H^n_{\mathbf{K}}$, $X = H^{n'}_{\mathbf{K}'}$, $\mathbf{K}(p,t)$ denotes a d_0-ball on X_0, and $\mathbf{K}'(p',t)$ denotes a d_0-ball on X'_0.

Given $p \in X_0$, set $\infty = -p$, let ρ denote the projection of X onto $X_0 - \infty$ along geodesic lines issuing from ∞ and let ρ' denote the corresponding projection of X' onto $X' - \phi(\infty)$.

By hypothesis, there are (k,b) pseudo-isometries $\phi : X \to X'$ and $\phi' : X' \to X$ which are Γ-space morphisms. By Lemma 14.1, there is a map $\overline{\phi}$ of geodesic lines in X to geodesic lines in X' such that for any geodesic line F in X

$$hd(\phi(F), \overline{\phi}(F)) < c \tag{21.11}$$

where c is a constant independent of F. We select a positive number r such that $r > b$ and $k^{-1}r - c > 0$. Set $R = d_X(v,o)$ for $v \in X$. We lose no generality in assuming $\phi(o) = 0$, since ϕ can be replaced if necessary by $\phi \circ g$ with g an isometry of X. Let F denote the geodesic line $[\infty,p]$, let v be a point of F such that the ray from v to p lies at a distance R from o, and let v' be the nearest point on $\overline{\phi}(F)$ to $\phi(v)$. Then $d_{X'}(v',\phi(v)) < c$ and we have the inclusions

$$\begin{aligned} B(v',k^{-1}r-c) &\subset B(\phi(v),k^{-1}r) \subset \phi(B(v,r)) \\ &\subset B(\phi(v),kr) \subset B(v',kr+c) \end{aligned} \tag{21.12}$$

as well as the inequalities

$$k^{-1}R - c \leq R' \leq kR + c$$

where $R' = d_{X'}(v',o)$. Clearly $\overline{\phi}(F)$ is the geodesic line with boundary points $\phi_0(\infty)$ and $\phi_0(p)$, by (21.11).

We can select R sufficiently large such that for any finite set of radii r with $r > b$ and $k^{-1}r - c > 0$,

$$\mathbf{K}(p,5^{-1}e^{-R+r}) \subset \rho(B(v,r)) \subset \mathbf{K}(p,2^{-1}e^{-R+r})$$

(21.13)

$$K'\left(\phi_0(p),5^{-1}e^{-R'+k^{-1}r-c}\right) \subset \rho'(B(v',k^{-1}r-c)\ ,$$

and

$$\rho'(v',kr+c) \subset \mathbf{K}'(\phi_0(p),2^{-1}e^{-R'+kr+c})\ ,$$

by Lemma 21.1. By definition of $\overline{\phi}$ and ϕ_0,

$$\phi_0 \circ \rho = \rho' \circ \phi_0\ . \tag{21.14}$$

Consequently

$$\begin{aligned} K'\left(\phi_0(p),5^{-1}e^{-R'+k^{-1}r-c}\right) &\subset \rho'(B(v',k^{-1}r-c) \\ &\subset \rho'(B(\phi(v),k^{-1}r) \\ &\subset \rho'(\phi(B(v,r)) \\ &\subset \phi_0(\rho(B(v,r)) \\ &\subset \phi_0(\mathbf{K}(p,2^{-1}e^{-R+r})\ . \end{aligned}$$

By (21.1), $\mathbf{K}(p,2^{-1}e^{-R+r}) \subset \rho(B(v,r+\log(5/2))$. Set

$$t = 2^{-1}e^{-R+r}\ . \tag{21.15}$$

Then

$$\begin{aligned} \phi_0(\mathbf{K}(p,t)) &\subset \phi_0(\rho(B(v,r+\log 5/2))) \\ &\subset \rho'(\phi(v,r+\log 5/2)) \\ &\subset \rho'(B(v',kr+c')) \\ &\subset \mathbf{K}'(\phi_0(p),2^{-1}e^{-R'+kr+c'}) \end{aligned}$$

where $c' = c + k\log(5/2)$.

Set

$$(21.16) \qquad \begin{aligned} s &= 5^{-1} e^{-R' + k^{-1} r - c} \\ \kappa &= 2^{-1} e^{kr - k^{-1} r + c' + c} \, . \end{aligned}$$

Then the inclusions above imply (21.10). The proof of Lemma 21.2 is now complete.

REMARK. For any $p \in X_0$ and $t > 0$, we set

$$(21.17) \qquad \begin{aligned} L(p,t) &= \inf\{s;\ \phi_0(\mathbf{K}(p,t) \subset \mathbf{K}'(\phi_0(p),s\} \\ \ell(p,t) &= \sup\{s;\ \mathbf{K}'(p,s) \subset \phi_0(\mathbf{K}(p,t))\} \, . \end{aligned}$$

Condition (21.10) is equivalent to

$$(21.18) \qquad \lim_{t \to 0} \sup \frac{L(p,t)}{\ell(p,t)} \leq \kappa \, .$$

If we were dealing here with standard balls rather than d_0-balls, condition (21.18) would be the condition for being κ-quasi-conformal. For example, in case $\mathbf{K} = \mathbf{K}' = \mathbf{R}$, Lemma 21.2 asserts that ϕ_0 is κ-quasi-conformal in the standard sense.

We are now at last prepared to prove that the boundary map is absolutely continuous on almost all $\mathbf{R}$-circles.

As in Section 20, we set $X_* = X_0 - S^{k(n-1)-1} - S^{k-1}$, that is the boundary $kn-1$ sphere with the points on the first axis of $\mathbf{K}^n$ and its orthogonal complement deleted. Let $\pi_{\mathbf{R}} : X_* \to Y$ be the fibering described in Section 20, its fibers are open quarter great $\mathbf{R}$-circles.

PROPOSITION 21.3. *The boundary map $\phi_0 X_0 \to X'_0$ is absolutely continuous on almost all fibers of $\pi_{\mathbf{R}}$ except possibly in the case $X = H^2_{\mathbf{R}}$. Moreover, $k = k'$.*

Proof. We have $X = H^n_{\mathbf{K}}$ and $X' = H^{n'}_{\mathbf{K}'}$ with $k = \dim_{\mathbf{R}} \mathbf{K}$, $k' = \dim_{\mathbf{R}} \mathbf{K}'$. Set $m = kn-1$. Let $\pi^{\mathbf{K}}$ denote the fibering of X_* (resp. 6) provided by great K-spheres (cf. (20.12)). Given $y \in Y$, we let $D(y,t)$ denote the polydisc with center y and radius t (cf. (20.16)).

Let X_1 be a compact subset of X_*. For any $y \in Y$ and for any $t > 0$, set $E(y) = X_1 \cap \pi_R^{-1}(y)$ and $E(y,t) = \{p \in X_0;\ d_0(p,q) \leq t$ for some $q \in E(y)\}$, where d_0 is the boundary semi-metric defined in (20.14). Set

$$\tau(y) = \limsup_{t \to 0} \frac{\mu_m(\phi_0(E(y,t))}{t^{m+k-2}} . \tag{21.19}$$

Then for almost all $y \in Y$,

$$\tau(y) < \infty . \tag{21.20}$$

For up to a factor bounded away from 0 and ∞, t^{m+k-2} is the $m-1$ dimensional Lebesgue measure of a polydisc $D(y,t)$ in Y. Let X_2 denote the closure of $T_t(X_1) \cap X_0$. Then for t small enough, $X_2 \subset X_*$. By (20.17) there is a positive constant c such that

$$D(\pi_R(p), c^{-1}t) \subset \pi_R(K(p,t)) \subset D(\pi_R(p), ct)$$

for all $p \in X_2$. Consequently $E(y,t) \subset \pi_R^{-1}(D(y,ct)) \cap X_2$ and $\mu_{m-1}(\pi_R(E)y,t)) \geq \mu_{m-1}(D(y,c^{-1}t) \geq c_1 t^{m+k-2}$ for some constant c_1. Thus

$$\tau(y) \leq \limsup_{t \to 0} \frac{\Phi(D(y,ct))}{\mu_{m-1}(D(y,ct))}$$

where for any closed set S in Y, $\Phi(S) = c_1^{-1}\mu_m(\phi_0(\pi_R^{-1}(S) \cap X_2))$. It follows now by Lemma 20.3 that $\tau(y) < \infty$ for almost all $y \in Y$.

Let y be a point in Y such that $\tau(y) < \infty$. We shall prove that ϕ_0 is absolutely continuous on the fiber $\pi_R^{-1}(y)$. In order to achieve this, we shall prove:

(21.21) *There is a constant* A *such that for any compact subset* $E \subset \pi_R^{-1}(y)$

$$\mu_1(\phi_0(E))^{m+k-1} \leq A\, \tau(y)\, \mu_1(E)^{m+k-2} .$$

Here μ_1 denotes 1-dimensional *Hausdorff measure*. We recall its definition. For any subset $E \subset S^m$ and for any positive real number a, we write

$$\wedge(E,a) = \inf_{\{U\}} \sum_U 2 \text{ radius } U$$

where $\{U\}$ is a denumerable cover of E by standard balls of radius not exceeding a.

Our method of proving (21.21) will establish the equality

$$(21.22) \qquad k = k' .$$

Let $L(p,t)$ and $\ell(p,t)$ be as in (21.17). We can assume without loss of generality in our proof of (21.21) that there is a positive number b satisfying for all $p \in E$

$$(21.23) \qquad L(p,t) \leq \kappa \ell(p,t) \quad \text{whenever} \quad 0 < t < b .$$

For let E_b be the set of $p \in E$ such that (21.23) is satisfied. Since E_b is compact and expands to E as $b \to 0$, the validity of (21.21) for E follows from its validity for E_b. Next we fix a positive number a (which ultimately will tend to zero).

We can select a positive c such that $L(p,c) < a$ for all $p \in E$. For by (21.23) and Lemma 20.2 (ii), we see that $L(p,t)$ does not exceed $2^{-1}\kappa$ times the standard diameter of the set $\phi_0(\mathbf{K}(p,t))$ for any $p \in E$. Thus we need only choose c so that the standard diameter of $\phi_0(\mathbf{K}(p,c)) < 2\kappa^{-1}a$ for all $p \in E$; this we can do by the uniform continuity of ϕ_0.

We can find a real number t, $0 < t < \inf(b,c)$ and points $p_1, p_2, \ldots, p_N$ in the set E such that the standard balls $T_t(p_i)$ cover E, $|p_i - p_j| \geq (|i-j|-1)t$ for all $i,j = 1,\ldots,N$, and $Nt \leq \mu_1(E) + a$. Namely, let F denote the fiber $\pi_{\mathbf{R}}^{-1}(y)$; F is an open quarter of a great $\mathbf{R}$-circle. Choose $t < \inf(b,c)$ so that $\mu_1(T_t(E) \cap F) < \mu_1(E) + a$, divide the arc F into equal disjoint half-open intervals of length t, select in each interval meeting E a point of E, and arrange these points $p_1, p_2, \ldots, p_N$ in order. Set

$$s_i = L(p_i,t) \qquad (i = 1, \ldots, N) .$$

Clearly $s_i < a (i = 1, \ldots, N)$.

From (21.23) we get

$$\kappa^{-1} s_i \leq \ell(p_i,t)$$

and thus by (21.17)

$$K'(\phi_0(p_i),\kappa^{-1} s_i) \subset \phi_0(K(p_i,t)) \subset K'(\phi_0(p_i),s_i) .$$

Set

$$B_i = T_{s_i}(\phi_0(p_i))$$

$$D_i = K'(\phi_0(p_i),s_i) \qquad D_i = K'(\phi_0(p_i),\kappa^{-1} s_i) .$$

Since the B^i form a cover of $\phi_0(E)$ by standard balls of radius less than a, we have

$$\Lambda(\phi_0(E),a) \leq \sum_i 2s_i .$$

By Hölder's inequality

$$\text{(21.24)} \qquad \Lambda(\phi_0(E),a)^{m+k'-1} \leq 2^{m+k'-1} N^{m+k'-2} \sum_i s_i^{m+k'-1} .$$

Now $\phi_0 : X_0 \to X'_0$ being a homeomorphism, we have $\dim X_0 = m = \dim X'_0$. Furthermore, we have

$$\text{(21.25)} \qquad c_1 s^{m+k'-1} < \mu_m(K'(q,s)) < c_2 s^{m+k'-1}$$

for all $q \epsilon X'_0$ and $s > 0$ where $c_1 > 0$. Thus (21.24) yields

$$\Lambda(\phi_0(E),a)^{m+k'-1} \leq 2^{m+k'-1} c_1^{-1} N^{m+k'-2} \sum_i \mu_m(D^i) .$$

Since $D_i \subset \phi_0(K(p_i,t))$, each point of D_i lies in at most three of $D_1, \ldots, D_N$. Hence

$$\sum_i \mu_m(D_i) \leq 3\mu_m\left(\bigcup_i D_i\right)$$

and

$$\sum_i \mu_m(D^i) \le c_2 \kappa^{m+k'-1} \sum_i \mu_m(D_i) \le 3c_2 K^{m+k'-1} \mu_m\Big(\bigcup_i D_i\Big)$$
$$\le 3c_2 \kappa^{m+k'-1} \mu_m\Big(\phi_0\Big(\bigcup_i K(p_i,t)\Big)\Big) .$$

Set $E(y,t) = \{p \in X_0;\ d_0(p,q) \le t$ for some $q \in E\}$. Since $K(p_i,t) \subset E(y,t)$ for each i, we deduce

$$\Lambda(\phi_0(E),a)^{m+k'-1} \le A\, N^{m+k'-2} \mu_m(\phi_0(E(y,t))$$

where $A = 3c_2\, c_1^{-1}\, (2\kappa)^{m+k'-2}$. Consequently

$$\text{(21.26)} \qquad \Lambda(\phi_0(E),a)^{m+k'-1} \le A(Nt)^{m+k'-2}\ \frac{\mu_m(\phi_0(E(y,t))}{t^{m+k-2}} \cdot \frac{t^{m+k-2}}{t^{m+k'-2}}$$
$$\le A(\mu_1(E) + a)^{m+k'-2}\, \tau(y)\, t^{k-k'}$$

for t sufficiently small.

We can certainly assume with no loss of generality that $k \ge k'$. If $k > k'$, we would deduce from (21.26) that $\Lambda(\phi_0(E,a)) = 0$ and thus $\mu_1(\phi_0(E)) = 0$ for every compact subset of $\pi_{\mathbf{R}}^{-1}(y)$ – contradicting the fact that the ϕ_0 is a homeomorphism. Consequently we deduce

$$\text{(21.27)} \qquad k = k' .$$

Now letting $a \to 0$ in (21.26), we infer assertion (21.21).

Suppose now that $X \neq H^2_{\mathbf{R}}$. Then $m = nk - 1 \ge 2$ and therefore $m + k' - 2 \ge 1$. It is clear now that if $\mu_1(E) = 0$, then $\mu_1(\phi_0(E)) = 0$. Therefore ϕ_0 is absolutely continuous on $\pi_{\mathbf{R}}^{-1}(y)$. The proof of Proposition 21.3 is now complete.

COROLLARY 21.4. *Assume* $nk > 1$. *The boundary map* ϕ_0 *is absolutely continuous on almost all* $\mathbf{R}$*-circles. In fact, given any* $\mathbf{K}$*-sphere* H *in* X_0 *and given any pair of distinct points* p,q *in* H, *the map* ϕ_0 *is absolutely continuous on almost all the* $\mathbf{R}$*-circles through* p *and* q, *for almost all* $(p,q) \in H \times H$.

Proof. By a transformation g from the group G, the point pair (p,q) can be sent into a pair of antipodal points (p,−p). The transformation g sends H into a great $\mathbf{K}$-sphere, and it sends the family of $\mathbf{R}$-circles through p and −p — the latter being great $\mathbf{R}$-circles. By Proposition 21.3, $\phi_0 g^{-1}$ is absolutely continuous on almost all the great $\mathbf{R}$-circles meeting gH. We can deduce Corollary 21.4 from this with the help of a direct product measure argument.

REMARK. Let S be an $\mathbf{R}$-circle on which ϕ_0 is absolutely continuous let ϕ_0^S denote the restriction of ϕ_0 to S let ϕ_p^S denote the differential of ϕ_0^S at a point $p \in S$ at which this differential exists. Then $\phi_p^S \neq 0$ for all p in a subset of positive measure in S. For let $p_0 \in S$ and let p(t) be a parametrization of S by arc length. Set $f(t) = \mu_1(\phi_0(p[0,t]))$ where $p[0,t] = \{p(s); 0 \leq s \leq t\}$. Then f is a strictly monotonic function, since ϕ_0 is a homeomorphism. Therefore, $\frac{df}{dt} \neq 0$ for t in a set of positive measure. From this, our assertion follows.

There is a striking consequence of Proposition 21.3 in another direction

COROLLARY 21.5. *Let* $G, G', X, X', \Gamma, \Gamma', \theta, \phi$ *be as above. Then* $G = G'$ *and* $X = X'$.

Proof. Since $k = k'$ and $nk = \dim X_0 + 1 = \dim X'_0 + 1 = n'k'$, we get $n = n'$. Thus $X = H^n_{\mathbf{K}} = X'$. Hence $G = G'$.

This corollary is of course an important step in proving rigidity of (G,Γ) in the $\mathbf{R}$-rank 1 case. But rigidity entails the stronger assertion that θ extends to an analytic automorphism of G.

REMARK. As pointed out in (21.18), Lemma 21.2 asserts in effect that the boundary map ϕ_0 is *quasi-conformal*. Our proof of this fact makes use of the fact that ϕ is a pseudo-isometric Γ-morphism. It would be of interest to deduce the existence and the quasi-conformality of the boundary map of more general quasi-conformal maps over $\mathbf{K}$.

§22. The Effect of Ergodicity

We continue the notation and hypotheses of Section 21. Thanks to Corollary 21.5, we have $X = X'$ and $G = G'$.

Let $^{\blacktriangleleft}F$ be a chamber in X; since $\mathbb{R}$-rank $X = 1$, $^{\blacktriangleleft}F$ is simply a geodesic ray. Let $^{\blacktriangleleft}A$ denote the set of polar elements in $G_{\blacktriangleleft F} = \{g \in G;\ g^{\blacktriangleleft}F \subset {}^{\blacktriangleleft}F\}$. By Mautner's Lemma (cf. Lemma 8.4) we know that A operates ergodically on G/Γ. We now investigate the consequences of this ergodicity for the boundary map ϕ_0.

It will be instructive to prove that ϕ is continuous on $X \cup X_0$, where ϕ is understood to map X_0 via ϕ_0. A proof of this fact can be extracted from our proof of Lemma 21.2, but that one would not suit our purpose. The proof we present in Lemma 22.1 is valid for the Satake-Furstenberg compactification of a symmetric space X of arbitrary $\mathbb{R}$-rank.

LEMMA 22.1. *Define* $\phi(p) = \phi_0(p)$ *for all* $p \in X_0$. *Then* $\phi : X \cup X_0 \to X \cup X_0$ *is continuous.*

Proof. For any subsets B and C in X let $B^{\blacktriangleleft}C$ denote the union of all chambers $^{\blacktriangleleft}F$ with origin in B and intersecting C together with the boundary points $[^{\blacktriangleleft}F]$. Let $p \in X_0$ and let L be a regular geodesic ray in X which has p as its boundary point. A base of neighborhoods of p in $X \cup X_0$ is given by the family $B(L,r,s)$ of subsets of the form

$$B(x,r) \,^{\blacktriangleleft}\, B(y,s) \tag{22.1}$$

where $x \in L, y \in L, r > 0$ and $s > 0$. By Lemma 14.1, there is a constant v (cf. (13.2.7)) such that ϕ maps any geodesic line to within a Hausdorff distance v of another. By choosing r and s sufficiently large, we can see that the base $B(L, r, s)$ goes into an equivalent base under the pseudo-isometry ϕ of X.

LEMMA 22.2. *Let* p, p_∞ *be distinct points of* X_0 *and set* $p' = \phi_0(p)$, $p'_\infty = \phi_0(p_\infty)$. *Let* F *denote the geodesic line with endpoints* p *and* p_∞, *let* $^{\blacktriangleleft}F$ *be a geodesic ray on* $^{\blacktriangleleft}F$ *with endpoint* p_∞, *and let* $^{\blacktriangleleft}A$ *be the set of polar elements in* $G_{^{\blacktriangleleft}F}$. *Let* F', $^{\blacktriangleleft}F'$, *and* $^{\blacktriangleleft}A'$ *denote the corresponding line, ray, and semi-group for* p', p'_∞. *Assume that* $^{\blacktriangleleft}A\Gamma$ *is dense in* G *in the Lie group topology. Then given any element* m *in the maximum compact subgroup of* $Z(^{\blacktriangleleft}A)$ (cf. 2.6 (iv)) *there are unbounded sequences* $\{a_n\}$ *in* $^{\blacktriangleleft}A$, $\{a'_n\}$ *in* $^{\blacktriangleleft}A'$, *and an element* m' *in the maximum compact subgroup of* $Z(^{\blacktriangleleft}A')$ *such that for all* $x \in X \cup X_0$

$$\phi(x) = m' \lim_{n\to\infty} a'_n \phi(a_n^{-1} m x) .$$

Proof. Inasmuch as G is doubly transitive on X_0, no generality is lost in assuming, after replacing ϕ by $\phi \circ g$ if necessary, that $p' = p$ and $p'_\infty = p_\infty$.

By hypothesis there is an unbounded sequence of elements $a_n \in {}^{\blacktriangleleft}A$ such that $a_n \gamma_n^{-1} \to m$ as $n \to \infty$, with $\gamma_n \in \Gamma$. Set $w_n = m^{-1} a_n \gamma_n^{-1}$; then $w_n \to 1$ as $n \to \infty$ and $\gamma_n = w_n^{-1} a_n m^{-1}$.

Consider now the map $\overline{\phi}$ sending geodesic lines of X to geodesic lines of X′. Let F denote the geodesic line in X with endpoints at p and p_∞. By Lemma 14.1 $\overline{\phi}$ is θ-equivariant and therefore

$$\phi(w_n^{-1} F) = \overline{\phi}(w_n^{-1} a_n m F) = \overline{\phi}(\gamma_n F) = \theta(\gamma_n)\overline{\phi}(F) = \theta(\gamma_n) F .$$

Inasmuch as $\overline{\phi}$ is a homeomorphism, we have $\overline{\phi}(w_n^{-1}F) = w'_n F$ with $w'_n \to 1$ as $n \to \infty$. Consequently

$$\theta(\gamma_n) = w'_n b'_n$$

with $b'_n \in G_F = Z(^{\blacktriangleleft}A)$. It follows that for all $x \in X$,

$$\phi(x) = \theta(\gamma_n)\phi(\gamma_n^{-1}x) = w'_n b'_n \phi(a_n^{-1} m w_n x) .$$

We define the map $\phi_n : X \cup X_0 \to X \cup X_0$ by

$$\phi_n(x) = b'_n \phi(a_n^{-1} m x) .$$

Then for all $x \epsilon X$

$$d_X(\phi(x), \phi_n(x)) \le d_X(\phi(x), w'^{-1}_n \phi(x)) + d_x(w'^{-1}_n \phi(x), \phi_n(x))$$

and

$$\begin{aligned} d_X(w'^{-1}_n \phi(x), \phi_n(x)) &\le d_X(\phi(a_n^{-1} m w_n x), \phi(a_n^{-1} m x)) \\ &\le k\, d_X(wnx, x) \end{aligned}$$

since ϕ is a (k,b) pseudo-isometry on X. Thus for all $x \epsilon X$,

(22.2) $$d_X(\phi(x), \phi_n(x)) \le d_X(w'_n \phi(x), \phi(x)) + k\, d_X(w_n x, x) .$$

It follows at once that $\phi_n(x) \to \phi(x)$ as $n \to \infty$ for all $x \epsilon X$ *uniformly* on compact subsets of X. We must have the stronger assertion

(22.3) $$\phi_n(x) \to \phi(x) \quad \text{for all} \quad x \epsilon X \cup X_0 .$$

In order to prove (22.3), we observe that the composition of an isometry with a (k,b) pseudo-isometry is a (k,b) pseudo-isometry and therefore each ϕ_n is a (k,b) pseudo-isometry. Hence for all n and for all geodesic lines F in X

$$hd(\phi_n(F), \overline{\phi}_n(F)) < v$$

where v is a constant (cf. (13.2.7)) depending only on X, k, and b, by Lemma 14.1.

Now let $q \epsilon X_0$ and let $B(L, r, s)$ be a base of neighborhoods of q in $X \cup X_0$ having the form (22.1); here L is a geodesic ray having q as an endpoint. Let L' be a geodesic ray such that $hd(\phi(L), L') < v$. Given any x and y in L, we have by (22.2)

$$\phi_n(B(x,r)) \subset B(\phi(x),\ kr + v)$$

$$\phi_n(B(y,s)) \subset B(\phi(y),\ kr + v)$$

for all $n > n_0$. From this it follows that there are points x' and y' in L' with

$$\phi_n(B(x,r)) \subset B(x',\ kr + 2v)$$

$$\phi_n(B(y,v)) \subset B(y',\ kr + 2v)$$

for $n > n_0$. It follows at once that $\phi_n(q) \to \phi(q)$ as $n \to \infty$. Thus (22.3) is now proved. As a consequence

$$\phi_0(q) = \lim_{n \to \infty} b'_n \phi_0(a_n^{-1} m q) \tag{22.4}$$

for all $q \in X_0$, where a_n is a sequence of elements in $^{\blacktriangleleft}A$ with $a_n \to \infty$ as $n \to \infty$, and $b'_n \in Z(^{\blacktriangleleft}A)$. Set $b'_n = m'_n a'_n$ with $a'_n = pol\, b'_n$. Replacing a_n by a subsequence, if necessary, we can assume that $m'_n \to m'$ as $n \to \infty$, since m'_n varies in the maximum compact subgroup of $Z(^{\blacktriangleleft}A)$.

Let A denote the set of polar elements in G_F. Each element a in A with $a \neq 1$ contracts neighborhoods of p_0 or p_∞ and expands neighborhoods of the other. It follows therefore from (22.4) that a'_n is contractive at p_∞ if a_n is. Therefore $a'_n \in {}^{\blacktriangleleft}A$ for all large n. By a similar argument we infer that $\{a'_n\}$ is unbounded. Lemma 22.2 is thus proved.

Before stating our next lemma, it will be convenient to introduce some new notation. Given two distinct points p, p_∞ on X_0, we denote by $\mathbf{R}\{p,p_\infty\}$ the union of all $\mathbf{R}$-circles passing through p and p_∞, and we let $\mathbf{K}[p,p_\infty]$ denote the $\mathbf{K}$-sphere passing through p and p_∞. Let N denote the unipotent radical of the stabilizer of the point p_∞, and let $\tau^p : N^p \to X_0$ denote the map $g \to gp$. Then, as is easily verified (cf. (19.12)), τ^p is a diffeomorphism of N onto $X_0 - p_\infty$. Let $\mathfrak{N}$ denote the Lie algebra of N, which we identify with the tangent space to N at the identity element, and let $\sigma^p = \tau^p \circ \exp$. Let $\mathfrak{N}_\alpha^p$ and $\mathfrak{N}_{2\alpha}^p$ denote the root spaces of the polar subgroup A^p in $\mathfrak{N}$, where A^p denotes the maximum polar subgroup of the stabilizer G_{F^p} of the geodesic F^p joining p and p_∞; that is, $A = pol\, F^p$. Then under the differential of σ^p at the origin, $\mathfrak{N}^p_\alpha$ and $\mathfrak{N}^p_{2\alpha}$ become identified with the tangent spaces to $\mathbf{R}\{p,p_\infty\}$ and $\mathbf{K}[p,p_\infty]$ at p. As remarked in (20.19), $\mathfrak{N}^p_\alpha$ is stable under the stabilizer of the point p. Furthermore $\mathfrak{N}^p_{2\alpha}$ is the center of $\mathfrak{N}$ and is independent of $p \in X_0 - p_\infty$.

We introduce in $\mathfrak{N}$ an operation corresponding to group multiplication in N via the formula

$$x \cdot y = \log(\exp x \cdot \exp y) = x + y + \frac{1}{2}[x,y] \ . \tag{22.8}$$

For any element g in the group G and for any distinct points p,q in X_0, we have

$$g\mathbf{R}\{p,q\} = \mathbf{R}\{gp,gq\} \ .$$

In particular, for $x \in \mathfrak{N}$ we have for any $p \in X_0 - p_\infty$, $\mathbf{R}\{(\exp x)\,p,p_\infty\} = (\exp x)\,\mathbf{R}\{p,p_\infty\}$. Thus for all $p \in X_0 - p_\infty$ and $x \in \mathfrak{N}$,

$$\sigma^p(x \cdot \mathfrak{N}^p_\alpha) = \mathbf{R}\{(\exp x)\,p,p_\infty\} \ . \tag{22.9}$$

The subset $x \cdot \mathfrak{N}^p_\alpha$ is an affine subspace of $\mathfrak{N}$ by (22.8). Similarly

$$\sigma^p(x \cdot \mathfrak{N}_{2\alpha}) = \mathbf{K}[(\exp x)\,p,p_\infty] \ .$$

The affine subspace $x \cdot \mathfrak{N}_{2\alpha}$ coincides with $x + \mathfrak{N}_{2\alpha}$.

Let F(p,q) denote the geodesic line in X whose endpoints in X_0 are p and q; let $^\blacktriangleleft F(p,q)$ be a geodesic ray in F(p,q) whose endpoint is q. Set $^\blacktriangleleft A(p,q) = pol\, ^\blacktriangleleft F(p,q)$ (see Section 7 for notation). By Lemma 8.4 (ii), $^\blacktriangleleft A(p,q)\Gamma$ is dense in G for almost all pairs $(p,q) \in X_0 \times X_0$. By Corollary 21.4, the boundary map is absolutely continuous on almost all $\mathbf{R}$-circles through p and q for almost all $(p,q) \in X_0 \times X_0$, provided that $nk > 1$. Finally, on any $\mathbf{R}$-circle S on which ϕ_0 is absolutely continuous, there is a subset of positive linear measure at which the differential of ϕ_0 along S exists and is non-zero (cf. Remark following Corollary 21.4).

For any $\mathbf{R}$-circle S, set

$$S_+ = \{q \in S;\ \dot{\phi}_q^S \text{ exists and is non-zero}\}\ .$$

For any $q \in X_0$, set

$$X\langle q\rangle = \{p \in X_0;\ {}^{\blacktriangleleft}A(p,q)\Gamma \text{ is dense in } G\}$$

and

$$X[q] = \{p \in X_0;\ \mu_1(S - X\langle q\rangle) = 0 \text{ and } \phi_0 \text{ is absolutely continuous on } S \text{ for almost all } \mathbf{R}\text{-circles } S \text{ through } p \text{ and } q\}\ .$$

From the three assertions above one infers at once:

(22.5) There exists a point p_∞ in X_0 such that $X\langle p_\infty\rangle$ and $X[p_\infty]$ differ from X_0 in a set of measure zero.

We now choose a point p_∞ satisfying (22.5). The point p_∞ will be fixed throughout this section. Set

$$X_+ = X\langle p_\infty\rangle \cap X[p_\infty]\ .$$

Then $X_+ \subset X_0 - p_\infty$ and X_+ differs from X in a set of measure zero.

LEMMA 22.3. *Assume* $\phi_0(p_\infty) = p_\infty$. *For any* $p \in X_0 - p_\infty$, *set*

$$\psi^p = (\sigma^p)^{-1}\ \phi_0\ \sigma^p\ .$$

Then for each $x \in \mathfrak{N}_a^p$, ψ^p *is an affine map on the line* $\mathbf{R}x$, *and* $\psi^p(\mathbf{R}x) \subset \psi^p(0) \cdot \mathfrak{N}_a^p$, *except possibly in the case* $X = H^2_{\mathbf{R}}$.

Proof. It suffices to prove Lemma 22.3 for all $p \in X_+$; for X_+ is dense in X_0 and the desired conclusion follows from the continuity of ϕ_0'

Let $p \in X_+$. Let S be an $\mathbf{R}$-circle through p and p_∞ such that (i) ϕ_0 is absolutely continuous on S and (ii) ${}^{\blacktriangleleft}A(q,p_\infty)\Gamma$ is dense in G for almost all $q \in S$. Since $p \in X_+$, almost all $\mathbf{R}$-circles through p and

p_∞ have the above property. For any **R**-circle S through p and p_∞, $(\sigma^p)^{-1}S = \mathbf{R}x$ with $x \in \mathfrak{N}_\alpha^p$ and conversely $\sigma^p(\mathbf{R}x) \cup p$ is an **R**-circle through p and p_∞ for any non-zero $x \in \mathfrak{N}_\alpha^p$. Therefore it suffices to prove the affineness of ψ^p on lines $(\sigma^p)^{-1}S$ for **R**-circles S through p and p_∞ satisfying (i) and (ii); for the desired conclusion will follow from the continuity of ψ^p. We are reduced therefore to proving the affineness of ψ^p on $\mathbf{R}x$, where $\sigma^p(\mathbf{R}x) \cup p_\infty$ satisfies (i) and (ii). Clearly no generality is lost in replacing ϕ by $g^{-1}\phi$ where g is the element of N such that $gp = \phi_0(p)$. Accordingly, we assume $\psi^p(o) = (o)$.

For simplicity of notation (since the point p will be fixed in the ensuing argument), we set $\mathfrak{N}_\alpha = \mathfrak{N}_\alpha^p$, $\sigma = \sigma^p$, $\psi = \psi^p$.

As remarked above, the subset S_+ has positive linear measure. Therefore by (ii) we can choose a point $p_0 \in S_+$ such that $^\blacktriangleleft A(p_0, p_\infty)\Gamma$ is dense in G. Set $x_0 = \sigma^{-1}(p_0)$ and $x'_0 = \psi(x_0)$. Then $x_0 = t_0 x$ with $t_0 \in \mathbf{R}$ and $(\exp x_0)p = p_0$.

Define for any $y \in \mathfrak{N}$ the maps $\psi_\alpha^y : \mathfrak{N} \to \mathfrak{N}_\alpha$ and $\psi_{2\alpha}^y : \mathfrak{N} \to \mathfrak{N}_{2\alpha}$ by the formula

$$\psi(y \cdot z) = \psi(y) \cdot \psi_\alpha^y(z) + \psi_{2\alpha}^y(z), \quad z \in \mathfrak{N} . \tag{22.6}$$

Now given any element $a \in {^\blacktriangleleft A}(p, p_\infty)$, $\sigma^{-1} a \sigma$ operates on $\mathfrak{N}_\alpha$ as multiplication by $\alpha(a)$ and on $\mathfrak{N}_{2\alpha}$ as multiplication by $\alpha(a)^2$. Inasmuch as $(\exp x_0)p = p_0$, we have $\exp x_0 \; {^\blacktriangleleft A}(p, p_\infty) \exp - x_0 = {^\blacktriangleleft A}(p_0, p_\infty)$ and $\sigma^{-1}(\exp x_0)a\,(\exp - x_0)^\sigma$ sends $x_0 \cdot y$ to $x_0 \cdot \alpha(a)y$ for any $a \in {^\blacktriangleleft A}(p, p_\infty)$ and $y \in \mathfrak{N}_\alpha$. Thus Lemma 22.2 asserts in effect:

There are sequences of positive real numbers t_n *and* t'_n *converging to zero such that for any* $y \in \mathbf{R}x$,

$$\psi(x_0 \cdot y) = \psi(x_0) \cdot m \lim_{n \to \infty} \left(t_n'^{-1} \psi_\alpha^{x_0}(t_n y) + t_n'^{-2} \psi_{2\alpha}^{x_0}(t_n y)\right) \tag{22.7}$$

where m *is an element in the maximum compact subgroup of* $Z({^\blacktriangleleft A}(p, p_\infty))$.
Inasmuch as x and x_0 commute, we have $x_0 \cdot tx = x_0 + tx$ for all $t \in \mathbf{R}$.
By (22.6)

$$\frac{d}{dt}\psi(x_0 + tx) = \psi(x_0) \cdot \frac{d}{dt}\psi_\alpha^{x_0}(tx) + \frac{d}{dt}\psi_{2\alpha}^{x_0}(tx) .$$

Set

$$\lambda_\alpha = \frac{d}{dt}\psi_\alpha^{x_0}(tx) \quad \text{at} \quad t = 0$$

$$\lambda_{2\alpha} = \frac{d}{dt}\psi_{2\alpha}^{x_0}(tx) \quad \text{at} \quad t = 0 .$$

Since $p_0 \in S_+$, both λ_α and $\lambda_{2\alpha}$ exist and at least one of them is nonzero.

We consider two cases.

Case 1. $\lambda_{2\alpha} \neq o$. Then for all $s \in \mathbf{R}$,

$$\lim_{n\to\infty} t_n'^{-1}\, \psi_{2\alpha}^{x_0}(t_n s\, x) = \lim_{n\to\infty} t_n'^{-2} t_n s\, \lambda_{2\alpha} = ds\, \lambda_{2\alpha}$$

where $\lim\limits_{n\to\infty} t_n'^{-2} t_n = d < \infty$. Hence

$$\lim_{n\to\infty} t_n'^{-1} t_n = 0 \quad \text{and by (22.7)}$$

(22.8) $$\psi(x_0 + sx) = \psi(x_0) \cdot m\Big(\lim_{n\to\infty} t_n'^{-1} t_n s\, \lambda_\alpha + ds\, \lambda_{2\alpha}\Big)$$

$$= \psi(x_0) \cdot m(ds\, \lambda_{2\alpha})$$

$$= \psi(x_0) + dsm\, \lambda_{2\alpha}$$

since $m\,\lambda_{2\alpha} \in \mathfrak{N}_{2\alpha}$ and $\mathfrak{N}_{2\alpha}$ is central in $\mathfrak{N}$. Since $x_0 = tx$, we get from (22.8) that $o = \psi(o) = \psi(x_0 - t_0 x) = \psi(x_0) - dt_0 m\, \lambda_{2\alpha}$. Hence $\psi(x_0) \in \mathfrak{N}_{2\alpha}$ and $\psi(\mathbf{R}x) \subset \mathfrak{N}_{2\alpha}$.

Case 2. $\lambda_\alpha \neq o$. Here by (22.7) and (22.6)

$$\lim_{n\to\infty} t_n'^{-1} t_n s\, \lambda_\alpha = \lim_{n\to\infty} t_n'^{-1}\, \psi_\alpha^{x_0}(t_n s\, x) = m^{-1}\, \psi_\alpha^{x_0}(sx) .$$

Therefore $\lim_{n\to\infty} t_n'^{-1} t_n = c \neq 0$ inasmuch as $\psi_\alpha^{x_0}(s\,x) \neq o$ for small s. Since by (22.7) and (22.6)

$$\lim_{n\to\infty} t_n'^{-2}\, \psi_{2\alpha}^{x_0}(t_n x)$$

equals $m^{-1}\, \psi_{2\alpha}^{x_0}(x)$, it is finite; it follows that

$$o = \lim_{n\to\infty} t_n'^{-1} \psi_{2\alpha}^{x_0}(t_n x) = \left(\lim_{n\to\infty} t_n'^{-1} t_n\right)\lambda_{2\alpha} = c\,\lambda_{2\alpha}$$

and therefore $\lambda_{2\alpha} = o$. Thus we have for all $s \in \mathbf{R}$

$$\psi(x\,x) = \psi(x_0)\cdot c\,s\,m\lambda_\alpha + \psi_{2\alpha}^{x_0}(sx) \tag{22.9}$$

$$\frac{d}{ds}\psi_{2\alpha}^{x_0}(sx) = o \quad \text{at} \quad s = 0\ . \tag{22.10}$$

From (22.1) we have

$$\psi(x_0)\cdot c\,s\,m\lambda_\alpha \equiv \psi(x_0) + c\,s\,m\lambda\alpha \;(\text{mod}\ \mathfrak{N}_{2\alpha})\ .$$

Consequently $\psi(sx) \equiv \psi(x_0) + c\,s\,m\lambda_\alpha \;(\text{mod}\ \mathfrak{N}_{2\alpha})$ and from this it follows that

$$(-\psi(y))\cdot\psi(y+sx) \equiv c\,s\,m\lambda_\alpha \;(\text{mod}\ \mathfrak{N}_{2\alpha})$$

for all $y \in \mathbf{R}x$. Consequently $\frac{d}{dt}\psi_\alpha^y(tx)_{t=0} \neq o$ for all $y \in \mathbf{R}x$. It follows from (22.10) that

$$\frac{d}{ds}\psi_{2\alpha}^{y}(sx) = o \tag{22.11}$$

for all y in $\mathbf{R}x$ such that $\sigma(t) \in X\langle p_\infty\rangle$; by condition (ii) on our choice of p, $\frac{d}{ds}\psi_{2\alpha}^y(sx) = o$ for almost all $y \in \mathbf{R}x$, and in particular for $y = o$. Similarly, (22.9) applies to $x_0 = o$ and thus we get, since $\psi(o) = o$

$$\psi(sx) = c\,s\,m\lambda_\alpha + \psi_{2\alpha}^{0}(sx)\ . \tag{22.12}$$

Condition (22.11) implies that for almost all $t \in \mathbf{R}$,

$$\frac{d}{dt}\psi(tx) \in \left(1 + \frac{1}{2}\,\mathrm{ad}\,\psi(tx)\right)\mathfrak{N}_\alpha$$

since $z\cdot y - z = z + y + \frac{1}{2}[z,y] - z = \left(\frac{1}{2} + \mathrm{ad}\, z\right)y$. By (22.12) $\psi(tx) \in \mathbf{R}\, m\lambda_\alpha + \mathfrak{N}_{2\alpha}$ for all $t \in \mathbf{R}$. Consequently $\mathrm{ad}\,\psi(tx)\,\mathfrak{N}_\alpha = o$ and

$$\frac{d}{dt}\psi(tx) \in \mathfrak{N}_\alpha$$

for almost all t. It follows at once that $\psi(tx) \in \mathfrak{N}_\alpha$ for all $t \in \mathbf{R}$ and $\psi^0_{2\alpha}(tx) = o$ for all t. Therefore for all $t \in \mathbf{R}$

$$\psi(tx) = c\,t\,m\lambda_\alpha\ . \tag{22.13}$$

We claim next that Case 1 cannot occur. For the homeomorphism ψ carries almost all lines through o in $\mathfrak{N}_\alpha$ into either $\mathfrak{N}_\alpha$ or $\mathfrak{N}_{2\alpha}$. Since $\dim \mathfrak{N}_{2\alpha} < \dim \mathfrak{N}_\alpha$, no subset of positive measure in $\mathfrak{N}_\alpha$ can be mapped by ψ into $\mathfrak{N}_{2\alpha}$. Hence ψ maps almost all lines through o in $\mathfrak{N}_\alpha$ linearly into $\mathfrak{N}_\alpha$. By continuity ψ maps all lines through o in $\mathfrak{N}_\alpha$ linearly into $\mathfrak{N}_\alpha$.

The proof of Lemma 22.3 is now complete.

REMARKS. In the exceptional case $X = H^2_{\mathbf{R}}$, our proof shows

(22.14) *If the boundary map* ϕ_0 *has a non-zero derivative at some point, then* ϕ_0 *is a Möbius transformation.*

For given $p \in X_0$, we can find a point $p_\infty \in X_0$ so that $\blacktriangleleft A(p,p_\infty)\Gamma$ is dense in G. Then by our reasoning in Case 2, ψ^p is affine on $\mathfrak{N}_\alpha$. Therefore ϕ_0 is a Möbius transformation.

In any case, Lemma 22.3 implies at once

(22.15) ϕ_0 *sends all* $\mathbf{R}$*-circles to* $\mathbf{R}$*-circles* .

For by Lemma 22.3, (22.15) holds for all $\mathbf{R}$-circles through p_∞ for almost all $p_\infty \in X_0$. (22.15) follows by continuity.

The analogue of (22.15) for $\mathbf{K}$-spheres is also an easy consequence of Lemma 22.3.

LEMMA 22.4. *ϕ_0 sends all* **K**-*spheres to* **K**-*spheres.*

Proof. We can assume of course that $\mathfrak{N}_{2a} \neq 0$, otherwise the result is trivial. By (19.14) we see that for any $x \in \mathfrak{N}_a$, $[x, \mathfrak{N}_a] = \mathfrak{N}_{2a}$, and hence for all $x \in \mathfrak{N}$

$$x \cdot \mathfrak{N}_a \cap \mathfrak{N}_a \text{ is empty if and only if } x \in \mathfrak{N}_{2a} . \tag{22.16}$$

By Lemma 22.3, for any $x \in \mathfrak{N}$

$$\psi(x \cdot \mathfrak{N}_a) = \psi(x) \cdot \mathfrak{N}_a . \tag{22.17}$$

It follows at once from (22.16) and (22.17) that

$$\psi(\mathfrak{N}_{2a}) = \mathfrak{N}_{2a} . \tag{22.18}$$

Since $\sigma(\mathfrak{N}_{2a}) \cup p_\infty = \mathbf{K}[p, p_\infty]$, we see that ϕ_0 carries the **K**-sphere through p and p_∞ into a **K**-sphere for almost all $(p, p_\infty) \in X_0 \times X_0$. The desired conclusion follows by the continuity of ϕ_0.

We note as a special case of the Lemma

$$\psi(x \cdot \mathfrak{N}_{2a}) = \psi(x) \cdot \mathfrak{N}_{2a}, \quad x \in \mathfrak{N} . \tag{22.19}$$

§23. R-Rank 1 Rigidity Proof Concluded

We continue the notation of the preceding section. In particular, $\psi(o) = o$.

LEMMA 23.1. *The map* ψ *is linear on* $\mathfrak{N}_\alpha$.

Proof. For any $x, y \in \mathfrak{N}$, we have

$$x \cdot y = x + y + \frac{1}{2}[x,y]$$

with $[x,y] = 2\,\mathrm{Im}(x,y)$ when $\mathfrak{N}$ identical with $\mathbf{K}^{n-1} + \mathrm{Im}\,\mathbf{K}$ as in (19.14). From this one infers at once that

$$[x, \mathfrak{N}_\alpha] = \mathfrak{N}_{2\alpha} \quad \text{for} \quad x \notin \mathfrak{N}_{2\alpha}$$

$$x \cdot \mathfrak{N}_\alpha = x + \left(1 + \frac{1}{2}\,\mathrm{ad}\,x\right)\mathfrak{N}_\alpha, \quad x \in \mathfrak{N} .$$

Set $\mathfrak{N}_\alpha^x = \left(1 + \frac{1}{2}\,\mathrm{ad}\,x\right)\mathfrak{N}_\alpha$. Then

$$\mathfrak{N}_\alpha^x + \mathfrak{N}_\alpha = \mathfrak{N}, \quad \text{if} \quad x \in \mathfrak{N}_{2\alpha} . \tag{23.1}$$

We fix, for the next two paragraphs, an element x not lying in $\mathfrak{N}_{2\alpha}$. Given any $y \in \mathfrak{N}$, we write $y = y' + y''$ with $y' \in \mathfrak{N}_\alpha$ and $y'' \in \mathfrak{N}_\alpha^x$. For any non-zero element $z \in \mathfrak{N}_{2\alpha}$, we have $z' \neq o$, otherwise $z = z'' \in \mathfrak{N}_\alpha^x$ and $\left(1 + \frac{1}{2}\,\mathrm{ad}\,x\right)^{-1} z \in \mathfrak{N}_\alpha$ which is impossible. It follows at once that the map $z \to z'$ is single valued. The map $z \to z'$ is clearly linear.

Consider next the intersection

$$\mathfrak{N}_\alpha \cap (z+x) \cdot \mathfrak{N}_\alpha = (z' + x') + \mathfrak{N}_\alpha \cap \mathfrak{N}_\alpha^x ; \tag{23.2}$$

it is a parallel translate of $\mathfrak{N}_\alpha \cap \mathfrak{N}_\alpha^x$. By (23.1)

$$\dim \mathfrak{N}_\alpha - \dim \mathfrak{N}_\alpha \cap \mathfrak{N}_\alpha^x = \dim \mathfrak{N}_{2\alpha} .$$

Therefore as z varies over $\mathfrak{N}_{2\alpha}$, the family of affine subspaces (23.2) yields all the parallel translates on $\mathfrak{N}_\alpha \cap \mathfrak{N}_\alpha^x$ in $\mathfrak{N}_\alpha$.

Furthermore, given any $x \in \mathfrak{N}_\alpha$, there exist elements $u_1, u_2, \ldots, u_{2n-3}$ such that

$$\text{(23.3)} \qquad Rx = \mathfrak{N}_\alpha \cap \mathfrak{N}_\alpha^{u_1} \cap \ldots \cap \mathfrak{N}_\alpha^{u_{2n-3}} .$$

In order to see (23.3), we identify $\mathfrak{N}$ with $\mathbf{K}^{n-1} + \operatorname{Im}\mathbf{K}$, $\mathbf{K} = \mathbf{C}, \mathbf{H}$, or $\mathbf{O}$. Then $\mathfrak{N}_\alpha = \mathbf{K}x_i + \mathbf{K}x_2 + \ldots + \mathbf{K}x_{n-1}$ and

$$\sum_i s_i x_i, \sum_j t_j x_j = \sum_i (s_i \bar{t}_i - t_i \bar{s}_i) .$$

No generality is lost in assuming that $x = x_{n-1}$. Choose an element $\sqrt{-1}$ in $\mathbf{K}$ whose square is -1. Then we can take for our $2n-3$ elements

$$\text{(23.4)} \qquad x_1, \sqrt{-1}\, x_1, \ldots, x_{n-2}, \sqrt{-1}\, x_{n-2}, x_{n-1} .$$

For it is clear that $\mathfrak{N}_\alpha \cap \mathfrak{N}_\alpha^x = \mathfrak{N}_\alpha \cap Z(x)$. Choosing our $2n-3$ elements as in (23.4) yields for the right side of (23.3), $\mathfrak{N}_\alpha \cap Z(u_1) \cap \ldots Z(u_{2n-3})$ and this is clearly $\mathbf{R}x_{n-1}$.

By Lemma 22.3, $\psi(x \cdot \mathfrak{N}_\alpha) = \psi(x) \cdot \mathfrak{N}_\alpha$ for all $x \in \mathfrak{N}$. It follows from (23.2) and (23.3) that ψ sends parallel lines in $\mathfrak{N}_\alpha$ to parallel lines. Consequently, ψ is linear on $\mathfrak{N}_\alpha$.

LEMMA 23.2. *The map* ψ *is affine on the affine subspace* $x \cdot \mathfrak{N}_\alpha$ *for all* $x \in \mathfrak{N}$.

Proof. Given $x \in \mathfrak{N}$, set $g = \exp{-x}$. Then replacing ϕ by $g\phi$ replaces ψ by $g\psi$ and reduces Lemma 23.2 to Lemma 23.1.

Before stating the next lemma, we recall that $p = \sigma^P(0)$, $p_\infty = X_0 - \sigma^P(\mathfrak{N})$. Set $F = F(p,p_\infty)$ and $A = pol\, F$. Then $Z(A) = G_p \cap G_{p\infty}$ and

$$(23.5) \qquad Z(A) = M \cdot A \quad \text{(direct)}$$

where M is the maximum compact subgroup of $Z(A)$ (cf. 2.6 (iv) and Lemma 5.2). Let ψ_a denote the restriction of ψ to $\mathfrak{N}_a$. Similarly, for any $g \in Z(A)$, let g_a denote the restriction of g to $\mathfrak{N}_a$.

(23.6) The map $g \to g_a$ of $Z(A)$ to $Z(A)_a$ is injective.

For if g_a is the identity, then g leaves fixed every element of $\mathfrak{N}_a$ and therefore of $\mathfrak{N} = \mathfrak{N}_a + [\mathfrak{N}_a, \mathfrak{N}_a]$. From (23.5) it is clear that all the elements of $Z(A)$ are semi-simple. Since g leaves fixed all the root vectors in the complexification $\mathfrak{N}_{\mathbf{C}}$, it leaves fixed all the vectors in an opposite nilpotent subalgebra $\mathfrak{N}^-$ and therefore fixes the subgroup generated by $\exp \mathfrak{N}$ and $\exp \mathfrak{N}^-$. Since G has no compact factors, this latter subgroup coincides with G. And since by hypothesis G has no center, this implies g is the identity element of G if g_a is the identity.

For any $g \in Z(A)$, we denote by $g_{\mathfrak{N}}$ the action of g on $\mathfrak{N}$. By (23.6), the map $g \to g_{\mathfrak{N}}$ is an injective map of $Z(A)$ to $Z(A)_{\mathfrak{N}}$.

LEMMA 23.3. $\psi_a Z(A)_a \psi_a^{-1} = Z(A)_a$.

Proof. By Lemma 22.2 and the invariance of $\mathfrak{N}_a$ under ψ, we have (cf. the remarks preceding 22.7): For any element $m \in M$, there is an $m' \in M$ and a positive scalar $c(m)$ such that

$$\psi(mx) = c(m)\, m'\psi(x)$$

for all $x \in \mathfrak{N}_a$. Hence

$$(23.7) \qquad \psi_a \circ m_a \circ \psi_a^{-1} = c(m)\, m'_a \; .$$

Inasmuch as M is compact, the determinant of each element of M_a is of Modulus 1. Taking the modulus of the determinant of each side of (23.7), we find $c(m) = 1$ for all $m \in M$. Thus

(23.8) $$\psi_\alpha M_\alpha \psi_\alpha^{-1} = M_\alpha .$$

Since ψ_α is linear, $\psi_\alpha \circ g_\alpha(x) = \psi_\alpha(\alpha(g)x) = \alpha(g)\psi_\alpha(x) = g_\alpha \circ \psi_\alpha(x)$ for all $x \in \mathfrak{N}_\alpha$ and $g \in A$; that is $\psi_\alpha \circ g_\alpha = g_\alpha \circ \psi_\alpha$ for $g \in A$. Our lemma from this and (23.8).

LEMMA 23.4. (i) *ψ is linear on $\mathfrak{N}$.*

(ii) *$\psi([x,y]) = [\psi(x), \psi(y)]$ for all $x,y \in \mathfrak{N}$.*

(iii) *$\psi Z(A)_{\mathfrak{N}} \psi^{-1} = Z(A)_{\mathfrak{N}}$.*

Proof. Given $x \in \mathfrak{N}_\alpha$ and $z \in \mathfrak{N}_{2\alpha}$, we have $\psi(z+x) = \psi(z \cdot x) \in \psi(z \cdot \mathfrak{N}_\alpha) = \psi(z) \cdot \mathfrak{N}_\alpha$ by (22.17). By (22.18), $\psi(z) \in \mathfrak{N}_{2\alpha}$. Hence $\psi(z+x) = \psi(z) + x'$ with $x' \in \mathfrak{N}_{2\alpha}$. By (22.19), $\psi(x+z) = \psi(x \cdot z) \in \psi(x) \cdot \mathfrak{N}_{2\alpha} = \psi(x) + \mathfrak{N}_{2\alpha}$. Hence there is an element $z' \in \mathfrak{N}_{2\alpha}$ such that

$$\psi(z) + z' = z' + \psi(x) .$$

Since $\mathfrak{N}_\alpha \cap \mathfrak{N}_{2\alpha} = (0)$, it follows that $\psi(z) = z'$ and $\psi(x) = x'$, and in particular

(23.9) $$\psi(x+z) = \psi(x) + \psi(z)$$

for all $x \in \mathfrak{N}_\alpha$, $z \in \mathfrak{N}_{2\alpha}$.

We show next that ψ is linear on $\mathfrak{N}_{2\alpha}$. By Lemma 23.3, we know that the map $y \to \psi(x \cdot y)$ is a linear map on $\mathfrak{N}_\alpha$. We have for any $x \in \mathfrak{N}_\alpha$

(23.10) $$\psi(x \cdot y) = \psi\left(x + y + \frac{1}{2}[x,y]\right) = \psi(x+y) + \psi\left(\frac{1}{2}[x,y]\right)$$

by (23.9). Now the map $y \to \frac{1}{2}[x,y]$ is a surjective linear map of $\mathfrak{N}_\alpha$ onto $\mathfrak{N}_{2\alpha}$ for any non-zero x in $\mathfrak{N}_\alpha$. Let x be a non-zero element in $\mathfrak{N}_\alpha$ and let $f : \mathfrak{N}_{2\alpha} \to \mathfrak{N}_\alpha$ be a linear map such that $\frac{1}{2} \operatorname{ad} x \circ f(z) = z$ for all $z \in \mathfrak{N}_{2\alpha}$. Then by (23.10)

$$\psi(z) = \psi(x \cdot f(z)) - \psi(x + f(z)) .$$

Since ψ is linear on $\mathfrak{N}_\alpha$, we see that $z \to \psi(z)$ is linear on $\mathfrak{N}_{2\alpha}$ and therefore on $\mathfrak{N}$ by Lemma 23.1.

To prove (ii), we observe that $\psi(x \cdot \mathfrak{N}_\alpha) = \psi(x) \cdot \mathfrak{N}_\alpha$ and thus to each $y \in \mathfrak{N}_\alpha$ there corresponds a uniquely defined element $y^x \in \mathfrak{N}_\alpha$ such that $\psi(x \cdot y) = \psi(x) \cdot y^x$. Thus $\psi\left(x+y+\frac{1}{2}[x,y]\right) = \psi(x) + y^x + \frac{1}{2}[y(x),y^x]$. This yields by (23.9)

$$\psi(x+y) = \psi(x) + y^x$$

$$\psi([x,y]) = [\psi(x),y^x]$$

for any x,y in $\mathfrak{N}_\alpha$. By Lemma 23.1, $y^x = \psi(y)$. Consequently $\psi([x,y]) = [\psi(x),\psi(y)]$ for all $x,y \in \mathfrak{N}_\alpha$, and hence for all $x,y \in \mathfrak{N}_\alpha + \mathfrak{N}_{2\alpha} = \mathfrak{N}$. Thus (ii) is proved. Assertion (iii) follows at once from (i), (ii) and Lemma 23.3.

We come at last to the goal at which we have been aiming. But first we must insert a remark on notation. For any $g \in G$ we denote by g_{X_0} the operation of g on X_0. The map $g \to g_{X_0}$ is a topological homomorphism of G onto a closed locally compact group G_{X_0} of homeomorphisms of X_0 which is in fact an isomorphism since its kernel is $\bigcap_{q \in X_0} G_q = (1)$, since G has no compact normal subgroup other than (1).

THEOREM 23.5. $$\phi_0 G_{X_0} \phi_0^{-1} = G_{X_0}.$$

Proof. We fix two distinct points p and p_∞ in X_0 and can assume without loss of generality that $\phi_0(p) = p$ and $\phi_0(p_\infty) = p_\infty$; for this can be arranged upon replacing ϕ by some $g\phi$ with $g \in G$ since G is doubly transitive on X_0. Set $P = G_{p_\infty}$. Then P is a parabolic subgroup of G and $P = Z(A) \cdot N$ (semi-direct) where $Z(A)$ is the stabilizer of the geodesic line $F = F(p,p_\infty)$ in X, $A = \mathit{pol}\, F$, and N is the unipotent radical of P. Let $\mathfrak{N}$ denote the Lie algebra of N, and let $\psi : \mathfrak{N} \to \mathfrak{N}$ denote the map defined by

$$\phi_0((\exp x)p) = (\exp \psi(x))p$$

for all $x \in \mathfrak{N}$. By Lemma 23.4, ψ is a Lie algebra automorphism of $\mathfrak{N}$. For any x,y in $\mathfrak{N}$

$$\begin{aligned}
\phi_0((\exp x \cdot \exp y)p) &= \left(\exp\left(x + y + \frac{1}{2}[x,y]\right)\right)p \\
&= \left(\exp\left(\psi(x) + \psi(y) + \frac{1}{2}[\psi(x),\psi(y)]\right)\right)p \\
&= (\exp \psi(x) \cdot \exp \psi(y))p \\
&= (\exp \psi(x))\phi_0((\exp y)p) .
\end{aligned}$$

Since $Np = X_0 - p_\infty$ and $\phi_0(p_\infty) = p_\infty$, we infer for all $x \in \mathfrak{N}$

$$\phi_0(\exp x)_{X_0}\phi_0^{-1} = (\exp \psi(x))_{X_0} . \tag{23.11}$$

By (23.11) and Lemma 23.4 (iii) we get

$$\phi_0 P_{X_0} \phi_0^{-1} = P_{X_0} . \tag{23.12}$$

Moreover, for all $\gamma \in \Gamma$ and $q \in X_0$ we have $\phi_0(rq) = \theta(\gamma)\phi_0(q)$. Hence

$$\phi_0 \Gamma_{X_0} \phi_0^{-1} = \theta(\Gamma)_{X_0} \subset G_{X_0} .$$

Consequently

$$\phi_0(\Gamma P)_{X_0} \phi_0^{-1} = \phi_0 \Gamma_{X_0} \phi_0^{-1} \cdot \phi_0 P_{X_0} \phi_0^{-1} \subset G_{X_0} .$$

By Lemma 8.5, ΓP is dense in G. Consequently

$$\phi_0 G_{X_0} \phi_0^{-1} = \phi_0 \overline{\Gamma P_{X_0}} \phi_0^{-1} \subset \overline{(G_{X_0})} = G_{X_0} . \tag{23.13}$$

COROLLARY 23.6. *Let* G *and* G' *be* R-*rank* 1 *groups having no center and no compact normal subgroups other than* (1), *let* Γ *and* Γ' *be discrete subgroups such that* G/Γ *and* G'/Γ' *have finite Haar measure, and let* $\theta : \Gamma \to \Gamma'$ *be an isomorphism. Let* X *and* X' *denote the associated symmetric spaces. Assume*

(*) *There exist pseudo-isometries* $\phi : X \to X'$ *and* $\phi' : X' \to X$ *which are* Γ_1*-space morphisms for some subgroup* Γ_1 *of finite index in* Γ. *Then there is an analytic isomorphism*

$$\overline{\theta} : G \to G'$$

such that θ *is the restriction of* $\overline{\theta}$ *to* Γ, *provided that* $G \neq PSL(2,R)$. *Moreover, if* G/Γ *and* G'/Γ' *are compact, assumption* (*) *is satisfied.*

Proof. By Corollary 21.5, we may assume that $G = G'$ and $X = X'$.

Let τ denote the canonical isomorphism $g \to g_{X_0}$ of G onto G_{X_0}. Since ϕ_0 is a Γ-morphism, we have for all $\gamma \in \Gamma_1$

$$\tau(\theta(\gamma)) = \phi_0 \circ \tau(\gamma) \circ \phi_0^{-1} .$$

Set

$$\overline{\theta}(g) = \tau^{-1}(\phi_0 \tau(\gamma) \phi_0^{-1})$$

for all $g \in G$. Then θ is a topological automorphism of G. By a well-known theorem of E. Cartan, $\overline{\theta}$ is an analytic automorphism. Moreover, $\overline{\theta}$ coincides with θ on Γ by the argument in the last paragraph of Section 17. The fact that (*) is satisfied if G/Γ and G'/Γ' are compact is given by Lemma 9.2.

REMARK. It is a fact that any automorphism of a parabolic subgroup of a centerless semi-simple analytic group G extends to a unique automorphism of G, and one can deduce (23.13) from (23.12) by means of this observation.

§24. Concluding Remarks

We can recapitulate our results as follows

THEOREM 24.1. *Let* G *and* G′ *be semi-simple analytic groups having no center and no compact factors, and let* Γ *and* Γ' *be discrete co-compact subgroups of* G *and* G′ *respectively. Let* $\theta : \Gamma \to \Gamma'$ *be an isomorphism. Then* θ *extends to an analytic isomorphism of* G *onto* G′, *provided only that there is no analytic homomorphism* π *of* G *onto* PSL(2,R) *with* $\pi(\Gamma)$ *discrete.*

Proof. This follows at once from Theorem 18.1 and Corollary 23.6.

An equivalent formulation in terms of locally symmetric spaces is

THEOREM 24.1′. *Let* Y *and* Y′ *be compact locally symmetric Riemannian spaces of non-positive sectional curvature. If* Y *and* Y′ *have isomorphic fundamental groups, then up to normalizing constants,* Y *and* Y′ *are isometric provided that* Y *has no closed one or two dimensional geodesic subspaces which are direct factors locally.*

Proof. Let X and X′ denote the simply connected covering spaces of Y and Y′ respectively; they are symmetric Riemannian spaces. Let G and G′ denote the connected component of the identity in the groups of isometries generated by the symmetrics in the points of X and X′. Then G is transitive on X, and $Y = \Gamma \backslash X$ where Γ is a discrete co-compact subgroup of G; and similarly $Y' = \Gamma' \backslash X'$. The decomposition $G = G_0 \times G_1 \times \cdots \times G_n$ into its center G_0 and all its non-abelian simple factors $G_1, \ldots, G_n$ corresponds to a decomposition $X = X_0 \times \ldots \times X_n$ into its flat factor X_0 and all its irreducible factors. The G-invariant metrics on X are unique up to scalar multiples on each of the factors $X_1, \ldots, X_n$, and a choice of an inner product on X_0. A non-flat factor X_i is two dimensional if and only if it has constant negative curvature and $G_i =$ PSL(2,R). The image of X_i in Y is a direct factor locally, and it is closed if and only if the projection of G on G_i sends Γ onto a discrete subgroup of G_i.

The hypothesis that Y has no one dimension geodesic subspace as a direct factor locally implies that X_0 reduces to a point. Actually we make the apparently weaker hypothesis that Y has no *closed* one dimensional local factor. But the two hypotheses are equivalent in view of the fact that $\Gamma \backslash \Gamma X_0$ is a closed flat toroid in Y; or what is equivalent ΓG_0 is a closed subgroup of G. To see this assertion, we note that the connected component of the identity in $\overline{\Gamma G_0}/G_0$ is invariant under Γ and hence under G by Lemma 8.6. On the other hand $[\overline{\Gamma G_0}, \overline{\Gamma G_0}] \subset \overline{[\Gamma G_0, \Gamma G_0]} \subset \overline{[\Gamma, \Gamma]} \subset \Gamma$, and thus the normal analytic subgroup $(\overline{\Gamma G_0})^{\text{conn}}$ has its commutator subgroup in G_0. It follows at once that $\overline{\Gamma G_0}$ has no semi-simple analytic subgroup and therefore $\overline{\Gamma G_0} = \Gamma G_0$. Then $\Gamma \backslash \Gamma X_0 = \Gamma \backslash \Gamma G_0$ is closed in the compact space $\Gamma \backslash G$ and is compact. Moreover, $\Gamma \backslash \Gamma G_0 = \Gamma \cap G_0 \backslash G_0$ and this is a toroid since G_0 is the vector group of translations on X_0. The hypothesis that Y has no one dimensional closed factors implies that $G_0 = (1)$. Thus Theorem 24.1′ follows from Theorem 24.1.

REMARK 1. Although our main result has been stated for co-compact lattices only, the fact that G/Γ is compact was used in an essential way *only in the proof of* Lemma 9.2 asserting the existence of a pseudo-isometric Γ-morphism ϕ from X to X′. Indeed, in anticipation of proving the existence of such a ϕ for general lattices, the hypothesis that G/Γ is compact has been replaced by the existence of the pseudo-isometry ϕ throughout our exposition. The existence of a pseudo-isometric Γ-morphism has been recently proved by Gopal Prasad for non-compact irreducible lattices Γ in semi-simple linear analytic groups G having an $\mathbf{R}$-rank 1 factor (cf. his forthcoming paper "Strong Rigidity of $\mathbf{Q}$-rank 1 Lattices," Inventiones Math.). Together with announced results of G. A. Margulis (to appear in Proceedings of the 1971 Soviet Summer School at Budapest), one has the

THEOREM 24.2. *Let* G *and* G′ *be semi-simple analytic groups without center or compact factors, and let* Γ *and* Γ' *be discrete subgroups of* G *and* G′ *respectively such that* G/Γ *and* G'/Γ' *have finite Haar measure. Let* $\theta:\Gamma\to\Gamma'$ *be an isomorphism. Then* θ *extends to an analytic isomorphism of* G *onto* G′, *provided only that there is no analytic homomorphism* π *of* G *onto* PSL(2,R) *with* $\pi(\Gamma)$ *discrete.*

REMARK 2. The hypothesis that G and G′ have no compact factor is obviously necessary. On the other hand, one can relax the hypothesis that G have no center, provided one weakens the conclusion on the extendibility of θ from Γ to G. Indeed one can easily deduce from Theorem 24.2

COROLLARY 24.2. *Let* G *and* G′ *be analytic semi-simple groups with finite centers and having no compact normal subgroup other than* (1). *Let* Γ *and* Γ' *be discrete subgroups such that* G/Γ *and* G'/Γ' *have finite measure. Let* $\theta:\Gamma\to\Gamma'$ *be an isomorphism. Then* G *and* G′ *have a common covering group. If moreover,* G *is a covering group of* G′, *then there is a subgroup of finite index* Γ_0 *in* Γ *such that the restriction of* θ *to* Γ_0 *extends to an analytic homomorphism of* G *onto* G′.

Examples of isomorphisms $\theta:\Gamma\to\Gamma'$ which do not extend to analytic homomorphisms of G to G′ can be obtained as follows. Take $G = G'$ and $\Gamma \neq [\Gamma,\Gamma]$. Let ζ be a non-trivial homomorphism of Γ into the center of G, and let $\Gamma_0 = \ker\zeta$. Let $\Gamma' = \{\gamma\zeta(\gamma);\ \gamma\in\Gamma\}$ and for each

$\gamma \in \Gamma$, set $\theta(\gamma) = \gamma\zeta(\gamma)$. Let θ_0 denote the identity map on Γ_0. Then any extension of θ_0 to a homomorphism of G must be the identity map on G since Γ_0 is Zariski-dense in G([2a]). Therefore θ is not extendible to G.

REMARK 3. The relation of our strong rigidity to the "deformation-rigidity" results mentioned in Section 1 can be described as follows.

Let Γ be a finitely generated group and G a Lie group. Let $R(\Gamma,G)$ denote the set of all homomorphisms of Γ into G topologized by the topology of pointwise convergence. Let $\theta \in R(\Gamma,G)$. One denotes by $H^1(\Gamma, \mathrm{Ad}\circ\theta)$ the cohomology of Γ with coefficients in the Lie algebra of G regarded as a Γ module under $\mathrm{Ad}\circ\theta$. One defines θ to be "*infinitesimally rigid*" if $H^1(\Gamma,\mathrm{Ad}\circ\theta) = 0$, and θ to be *locally rigid* if the orbit of θ under inner automorphisms of G is open in $R(\Gamma,G)$. According to a theorem of A. Weil ([21c]) if θ is infinitesimally rigid, then θ is locally rigid. The infinitesimal rigidity of Γ in a semi-simple group G without compact or three dimensional factors was proved by A. Weil (cf. [21]) if Γ is a discrete co-compact subgroup of G, and by A. Borel and M. S. Raghunathan if Γ is arithmetic in G; in these cases (cf. [2c]) Borel has proved that $R(\Gamma,G)$ has only finitely many connected components and that $R_*(\Gamma,G)$ is finite modulo inner automorphisms of G, where $R_*(\Gamma,G)$ is the subset of $\theta \in R(\Gamma,G)$ such that θ is injective and $\theta(\Gamma)$ is a lattice in G. Strong rigidity of Γ in G implies that $R_*(\Gamma,G)$ consists of a *single point* modulo automorphisms of G.

REMARK 4. In [12a] I have proved that compact solvmanifolds are uniquely determined up to homeomorphism – in fact diffeomorphism, by their fundamental groups. Theorem 24.2 on strong rigidity asserts that G/Γ is uniquely determined by Γ if Γ is a lattice in G and G is a semi-simple group having no center, no compact factor, and no factor PSL(2,R). Inasmuch as two Riemann surfaces having the same fundamental group are diffeomorphic, one is led to the following

THEOREM. *Let* G *and* G′ *be simply connected analytic groups having no compact semi-direct factors. Let* Γ *and* Γ' *be lattices in* G *and* G′ *respectively. Let* $\theta : \Gamma \to \Gamma'$ *be an isomorphism. Then*

(1) *There is a subgroup* Γ_0 *of finite index in* Γ *and a diffeomorphism* $\phi : G/\Gamma_0 \to G/\theta(\Gamma_0)$.

(2) *Identifying* $\pi_1(G/\Gamma_0)$ *with* Γ_0, *the isomorphism of* Γ_0 *induced by* ϕ *coincides with* θ *on* $[\Gamma_0, \Gamma_0]\Delta$ *where* Δ *is the maximum normal solvable subgroup of* Γ_0.

The proof of this result will appear elsewhere.

Bibliography

[1]. Bass, H., Milnor, J., and Serre, J. P., Solution of the congruence subgroup program for $SL_n (n \geq 3)$ and $Sp_{2n} (n \geq 2)$, Publ. IHES, N° 33 (1967), pp. 59-137.

[2]. Borel, A., (a) Density properties for certain subgroups of semi-simple groups without compact components, Ann. of Math., vol. 72 (1960), pp. 179-188.

(b) Introduction aux groupes arithmetiques, Hermann, Paris, 1969.

(c) On the automorphisms of certain subgroups of semi-simple Lie groups, Proc. Bombay Colloquium on Algebraic Geometry, 1968, pp. 43-73.

[3]. Calabi, E., (a) On compact Riemannian manifolds with constant curvature I, Differential geometry, Proc. Sympos. Pure Math. vol. 3, Amer. Math. Soc., Providence, R.I. 1961, pp. 155-180.

(b) and Vesentini, E., On compact locally symmetric Kahler manifolds, Ann. of Math., vol. 71 (1960), pp. 472-507.

[4]. Chevalley, C., Théorie des groupes de Lie, t 2 Hermann, Paris, 1951.

[5]. Freudenthal, H., (a) Oktaven, Ausnahmegruppen und Oktavengeometrie, Math. Inst. Rijksuniv., Utrecht, 1951.

(b) Zur ebene Oktavengeometrie, Proc. Kon. Nederl. Akad. Wetensch. Ser. A. vol. 56 (1953), pp. 195-200.

[6]. Furstenberg, H., (a) A Poisson formula for semi-simple Lie groups, Ann. of Math., vol. 77 (1963), pp. 335-386.

(b) Poisson boundaries and envelopes of discrete groups Bull. Amer. Math. Soc. vol. 73 (1967), pp. 350-356.

[7]. Helgason, S., Differential Geometry and Symmetric Spaces, Academic Press, New York (1962).

[8]. Hochschild, G., (a) The automorphism group of a Lie group, Trans. Amer. Math. Soc. vol. 72 (1952), pp. 209-216.

(b) The structure of Lie groups, Holden Day, San Francisco (1965).

[9]. Jacobson, N., Lie Algebras, Interscience, New York (1962).

[10]. Margulis, G. A., On the arithmeticity of discrete groups. Soviet Math. Dokl. (translation) vol. 10 (1969), pp. 900-902.

[11]. Mautner, F. I., Geodesic flows on symmetric Riemann spaces, Ann. of Math. vol. 65 (1957), pp. 416-431.

[12]. Mostow, G. D., (a) Factor spaces of solvable groups, Ann. of Math. vol. 60 (1954), pp. 1-27.

(b) Some new decompositions of semi-simple groups, Memoirs Amer. Math. Soc. vol. 14 (1955), pp. 31-54.

(c) Self adjoint group, Ann. of Math. vol. 62 (1955), pp. 44-55.

(d) On maximal subgroups of real Lie groups, Ann. of Math. vol. 74 (1961), pp. 503-517.

(e) Homogeneous spaces of finite invariant measure. Ann. of Math. vol. 75 (1962), pp. 17-37.

(f) On the conjugacy of subgroups of semi-simple groups, Proc. of Symposia in Pure Math. vol. 9 (1966), pp. 413-419.

(g) Quasi-conformal mappings in n-space and the rigidity of hyperbolic space forms, Publ. IHES, vol. 34, pp. 53-104.

(h) Intersections of discrete subgroups with Cartan subgroups, Journal of Indian Math. Soc., vol. 34 (1970), pp. 203-214.

(i) Lectures on Discrete Subgroups, Tata Institute of Fundamental Research, 1970.

(j) The rigidity of locally symmetric spaces, Proc. Int. Congress of Math., 1970 vol. 2, pp. 187-197.

[13]. Prasad, G., and Raghunathan, M. S., Some structural properties of lattices in semi-simple groups, Ann. of Math., vol. 96 (1972), pp. 296-317.

[14]. Raghunathan, M. S., Cohomology of arithmetic subgroups of algebraic groups I, II, Ann. of Math. vol. 86 (1967), pp. 409-424, vol. 87 (1968), pp. 279-304.

[15]. Satake, I., On representations and compactifications of symmetric Riemannian spaces, Ann. of Math., vol. 71 (1960), pp. 77-110.

[16]. Selberg, A., On discontinuous groups in higher dimensional symmetric spaces, International Colloquium on Function Theory, Tata Institute of Fund, Research (Bombay) 1960, pp. 147-164.

[17]. Steenrod, N., The Topology of Fiber Bundles, Princeton Univ. Press, (1951).

[18]. Springer, T. A.,(a) On a class of Jordan algebras, Proc. Kon. Neder Akad. Wetensch. A, vol. 62 (1959), pp. 254-264.

(b) The projective octave plane I, II, Ibid, vol. 63 (1960), pp. 74-101.

[19]. Springer, T. A., and Veldkamp, F. D., Elliptic and hyperbolic octave planes I, II, III, Ibid, vol. 66, pp. 413-451 (1963).

[20]. Tits, J., Lectures on buildings of spherical type and finite B-N pairs, Springer Lecture Notes (to appear).

[21]. Weil, A., (a) On discrete subgroups of Lie groups, Ann. of Math. vol. 72 (1960), pp. 369-384.

(b) On discrete subgroups of Lie groups II, Ibid, vol. 75 (1962), pp. 578-602.

(c) Remarks on Cohomology of groups, Ibid, vol. 80 (1964), pp. 149-157.

[22]. Wolf, J. A., Discrete groups, symmetric spaces, and global holonomy, Amer. J. Math. vol. 84 (1962), pp. 527-542.

Library of Congress Cataloging in Publication Data

Mostow, George D.
Strong rigidity of locally symmetric spaces.

(Annals of mathematics studies, no. 78)
Bibliography: p.
1. Riemannian manifolds. 2. Symmetric spaces.
3. Lie groups. I. Title. II. Series.
QA649.M78 516'.36 73-13003
ISBN 0-691-08136-0